Passive Solar Energy in Buildings
Watt Committee Report Number 17

Members of The Watt Committee on Energy Working Group on Passive Solar Building Design

Prof. P.E. O'Sullivan, *University of Wales Institute of Science and Technology, Cardiff (Chairman)*

Dr N. Baker, *Energy Conscious Design, London*
Dr D.M.L. Bartholomew, *Energy Technology Support Unit, Harwell*
B. Berrett, *Royal Town Planning Institute*
E.G. Bevan, *Department of Energy, London*
Prof. B.J. Brinkworth, *University College of Wales, Cardiff*
Dr J.R. Britten, *Building Research Establishment, Garston*
D.A. Button, *Pilkington Glass Ltd, St Helens*
J. Campbell, *Ove Arup Partnership, London*
Dr I. Cooper, *Eclipse Research Consultants*
Dr V.H.C. Crisp, *Chartered Institution of Building Services Engineers*
D.M. Curtis, *Essex County Council, Chelmsford*
J. Doggart, *Energy Conscious Design, London*
M.E. Finbow, *Royal Institute of British Architects*
Dr D. Fisk, *Building Research Establishment, Garston*
R.W. Ford, *Institute of Ceramics*
D. Goodenough, *ARC Concrete Ltd, Bristol*
Dr D. Hawkes, *Martin Centre, Cambridge*
E.R. Hitchin, *Institution of Gas Engineers*
G.K. Jackson, *Pilkington Glass Ltd, St Helens*
E.J. Keeble, *Building Research Establishment, Garston*
Dr D. Lindley, *Taylor Woodrow, Greenford*
Dr J. Littler, *Polytechnic of Central London*
D.M. Lush, *Ove Arup Partnership, London*
J.L. Meikle, *Davis, Langdon & Everest, London*
N.O. Milbank, *Building Research Establishment, Garston*
Prof. J.K. Page, *University of Sheffield*
W.B. Pascall, *Royal Institute of British Architects*
M.J. Patel, *Dept of Education and Science, London*
P.H. Pitt, *Harrow*
D. Poole, *Hampshire County Council*
Dr R. Wensley, *London Business School*
H.W.H. West, *Institute of Ceramics*

Note. The Working Group appointed four sub-groups whose members are listed on pages 5, 23, 35 and 47.

Acknowledgements

The Watt Committee Working Group on Passive Solar Building Design is indebted to many individuals and organisations in the United Kingdom from whom information and comments were obtained in the course of this project.

The Watt Committee on Energy acknowledges with thanks financial assistance by the Department of Energy, which helped to defray the costs of the proceedings of the Working Group.

Passive Solar Energy in Buildings

Edited by

PATRICK O'SULLIVAN

Welsh School of Architecture, UWIST, Cardiff
Chairman of the Working Group on Passive Solar Building Design appointed by the Watt Committee on Energy

Report Number 17

Published on behalf of
THE WATT COMMITTEE ON ENERGY
by

LONDON AND NEW YORK

First published 1988 by Taylor & Francis

Published 2019 by Routledge
2 Park Square, Milton Park, Abingdon, Oxon OX14 4RN
52 Vanderbilt Avenue, New York, NY 10017

Routledge is an imprint of the Taylor & Francis Group, an informa business

First issued in paperback 2019

ISBN 13: 978-0-367-45071-7 (pbk)
ISBN 13: 978-1-85166-280-7 (hbk)

British Library Cataloguing in Publication Data

The Watt Committee on Energy

Working Group on Passive Solar Building Design.
Passive Solar Energy in Buildings—(Report: VNO 17)
1. Buildings. Designs and Construction
   Implications of Solar Energy.
I. Title. II. O'Sullivan, Patrick.
III. Series.
721'.0467

Library of Congress Cataloging-in-Publication Data

Passive solar energy in buildings.

(Watt Committee report; no. 17)
Bibliography: p.
Includes index.
1. Solar buildings—Design and construction.
2. Solar energy—Passive systems. I. O'Sullivan,
Patrick Edmund. II. Watt Committee on Energy.
III. Series.
TH7413.P3713 1988   720'.47   88-21467
ISBN 1-85166-280-4

Foreword

Although the range of professional disciplines represented on The Watt Committee on Energy is so wide, many of them fall into a number of readily identified groups — civil and mechanical engineering, basic science, resources, medicine and health, management and accounting, education, and last but not least of this list, architecture and building. This is a field that is of direct personal interest to everyone, embracing many kinds of professional activity, yet many aspects of it are not widely understood, especially if they involve new applications of old technology or the development of new technology for traditional purposes. That, in a nutshell, is why the use of passive solar techniques was selected by the Watt Committee Executive, after consultation with the Department of Energy, as one of the topics for study by a series of Working Groups.

In one sense, passive solar design has been in use almost ever since human beings began to provide themselves with permanent dwellings. Everyone knows that the simple dwellings of agricultural communities are often orientated to make use of sunlight and that, in an ordinary modern dwelling house, the addition of a conservatory can have a remarkable effect on internal temperature. The application of the passive solar principle to more sophisticated modern buildings goes much further, however, as the sections on schools and office buildings in this Report demonstrate. Perhaps this is no more than a matter of good design; but if so, in this respect at any rate, the standard of design is often not what it might be. Do other considerations outweigh the possible solar gain, or have some builders and their customers, in an age of plentiful energy supply, albeit relying mainly on dwindling fossil fuel resources, simply ignored this option?

In all the differing circumstances of life today, there is no one simple answer; but it is surely all to the good that the potential for passive solar design should be clearly stated and authoritatively assessed. This is the task that was undertaken by Professor Patrick O'Sullivan and the Working Group of which he is Chairman. On behalf of the Watt Committee Executive, I am pleased to have this opportunity of thanking them for the time and trouble that has gone into this Report and of commending it, both to architects, builders and town-planners, who have the responsibility for putting passive solar design into practice, and to industrialists, home-owners, educationalists, banks, building societies and others who, as customers, may benefit.

G.K.C. PARDOE
Chairman, The Watt Committee on Energy

Terms of Reference

When the Watt Committee Executive decided to appoint the Working Group, its main purpose was to improve public awareness of passive solar building design. This was seen as a contribution to one of the central objectives of The Watt Committee on Energy, as provided by its constitution, namely 'to disseminate knowledge concerning energy for the benefit of the public at large'.

A paper developing the implications of these objectives was prepared by Professor J.K. Page. After initial consideration of this paper by a steering group, Professor P.E. O'Sullivan accepted the Chairmanship of the full Working Group, which met for the first time in January 1985.

The Working Group was asked by the Watt Committee Executive to make an initial informal report to the Eighteenth Consultative Council meeting of The Watt Committee on Energy on 23 October 1985, leading to the publication of the present Report. The papers published here have been revised in the light of discussion at the Consultative Council meeting and of further consideration by the Working Group and its sub-groups.

Contents

Section 1

Passive Solar Techniques and their Application in Existing and New Technology

Patrick O'Sullivan

Director of Architectural Research and Development, Welsh School of Architecture, UWIST, and Chairman of the Working Group on Passive Solar Building Design

This paper is based on the introductory address to the Eighteenth Consultative Council Meeting of the Watt Committee on Energy, held in London on 23 October 1985.

From the first meeting of the Watt Committee Working Group on Passive Solar Building Design, its members have been aware of a dual role. The simplest definition of the task that the Watt Committee Executive had asked it to perform is that it should review current understanding and practice of passive solar design to see what stage had been reached in its application in buildings. Second, however, by implication, this enabled the Working Group to form a view about the prospects for the wider adoption of passive solar techniques in the United Kingdom and to suggest what action, if any, would be likely to promote them effectively.

That the group took a generally positive attitude to the prospects for passive solar design will be evident from the papers in this Report, although their principal objective is to consider the sources of information.

1.1 PASSIVE SOLAR DEFINITION

The Working Group devoted a good deal of time to considering what is meant by 'passive solar'. For the purposes of this Report, it is argued that passive solar design is the use of the form and fabric of the building to admit, store and distribute primarily solar energy for heating and lighting. Moreover, this passive design process, as well as saving fuel, frequently enhances the amenity of the building, often at no extra construction or maintenance costs. Passive solar concepts and their complementary energy conservation measures, when used together in energy-efficient buildings, offer a major opportunity for the reduction of dependence on the world's fossil fuel supplies at lower costs in industry, commerce and the home.

So argues the conventional wisdom, and moreover it ought to be true. Certainly, in our high-density urban environment, where the major energy requirement is for enormous quantities of low-grade heating for spaces and hot water, 'passive solar' is probably the one form of renewable energy that can make a major and/or measurable contribution in the foreseeable future.

1.2 COST VERSUS BENEFIT

Yet, although such ideas, desires and wishes have been around for a long time, for some reason, until now, there have not been many ordinary buildings that were designed and built explicitly to optimise such parameters. (There have, of course, been some notable examples and exceptions, such as the buildings at Basildon New Town, Milton Keynes,

and elsewhere, and a number of distinguished 'one-off' designs.) The reason seems to arise from the perpetual question of 'cost versus benefit'.

It may well be that the comparative simplicity of passive design has been misunderstood, and that consequently it has been thought that it is costly and that the benefits are uncertain: it has been tinged with the great fear of the 1960s — overheating. Moreover, the overall benefit has been difficult to assess — after all, the sun shines on all houses, and the question of how much more you get for doing things correctly has never, so far, been fully and satisfactorily explained.

1.3 EXISTING KNOWLEDGE

The purpose, then, of the Eighteenth Consultative Council meeting of the Watt Committee and of this Report is not to pretend that we have all the answers or that no further work needs to be done, or is indeed being done; it is to gather together, and to put before the public in a coherent and easily understandable form, what is jointly and separately known in professional circles and what we have been able to find out. Equally, and indeed more importantly, we hope to benefit from others' knowledge and experience. In the course of our studies we have been able to put together a statement of what is known about passive solar design in the form of an argument which, we believe, should be sufficient to convince the building industry of the benefits of this method of design and, more importantly, to persuade it to take some action. Again, it must be emphasised that this does not mean that we believe that in the United Kingdom we have all the answers, or that no further work is necessary; it means, rather, that we do know sufficient to make a start — and a sensible start at that. Current research and development promoted by the Department of Energy and the Department of the Environment, and other recent work, encourage us in this belief.

In 1985 the Working Group established four sub-groups to consider passive solar design in housing, offices, educational buildings and the industrial and retail sector. This Report consists of papers by the Chairmen of these groups, with concluding recommendations.

1.4 NEEDS AND OPPORTUNITIES

In general terms, the Working Group arrived at the following views:

(1) Owners and designers of buildings can choose whether to design them to be (a) climatically rejecting, i.e. exclusive, or (b) climatically accepting, i.e. interactive. Moreover, buildings in some sectors have more potential to be interactive than those in other sectors. For example, in industrial and commercial buildings the basic need is for daylight, not for heat: the building form should deal with this, either by means of natural ventilation, via courtyards or atria, or by ventilating deeper buildings mechanically.

(2) Location and the optimisation of site are economic facts of life. This means that, on any complex site, there should be mixed designs, some of which will be more interactive than others.

(3) The fact must be faced that in British winter conditions, heating must be provided by good, competent, efficient heating systems. The main benefits of passive solar are in spring and autumn; they include both the reduction of the quantity of heating energy needed and indeed the shortening of the heat season at both ends.

(4) There is plenty of solar energy about, even in Britain, and it is not too difficult to get it into the building. The major difficulty is in making sensible use of it once it is within the building fabric. Generally speaking, the best advantage comes from the pre-heating of spaces, such as those in hotels and domestic premises.

(5) It is new designs that must provide exemplars and show the consumer that there is also potential in the retrofitting of passive solar design to existing stock.

(6) It is argued in the Report that one of the main characteristics of passive solar buildings is their high amenity value. People like them and find them light, airy and 'joyful'. The expectation is that this amenity value will reflect itself in low occupant stress response that will in turn result in healthier people.

Section 2

Opportunities for Use of Passive Solar Energy in Educational Buildings

David Curtis

Energy Manager, Essex County Council, Chelmsford, Essex

This paper presents the work of a sub-group, dealing with schools and college buildings, of the Watt Committee Working Group on Passive Solar Building Design.

Membership of Sub-group

D. M. Curtis (Chairman)

D. A. Button
Dr I. Cooper
J. Doggart
G. K. Jackson
Dr D. Lindley
M. J. Patel

2.1 POLICY OF LOCAL EDUCATION AUTHORITIES

In the late nineteenth century, Robson, consultant architect to the Education Department at White-hall, commented: 'It is well known that the rays of the sun have a beneficial influence on the air of a room, tending to promote ventilation, and that they are to a young child very much what they are to a flower'.

There are currently 521 local authorities in the United Kingdom which consume about 20 000 000 tonnes of coal equivalent (20 mtce) energy per year. This equates to over £1 000 000 000 per annum in monetary terms, representing about 6% of the nation's energy consumption.

This energy is consumed in roughly the following proportions: 75% in buildings, 15% in transport and 10% in street lighting. In the school systems of Essex County Council, one of the largest local authorities but otherwise similar to other shire counties, approximately 800 schools use over 80% of the energy consumed in buildings. Equally good examples could be quoted from the areas of other local education authorities, but it was accepted in the Watt Committee sub-group on educational buildings that examples from Essex would be representative. It is important, therefore, to consider the potential impact, the technical state of the art, the barriers that must be overcome and the institutional changes that may have to be made to ensure that school building can benefit rather than suffer from passive solar energy. The construction and use requirements of schools favour the use of passive solar. Short daytime occupancy combined with high glazing in many existing schools and a requirement for a 'view' in newly built schools provide an excellent scenario for a calculated balance of various factors—daylighting, free heat in winter, high insulation of the building envelope and summer overheating.

Being the responsibility of a single authority in each county, schools lend themselves to replication of design and central control. In particular, the increasing use of electronic energy management systems, which often respond rapidly through compensation to external climatic changes, provides the other necessary factor—a quick controlled reduction in conventional heating to take maximum advantage of free solar heat. The purpose of passive design is to ensure that the form, fabric and systems of a building are arranged and integrated in order to maximise the benefits of ambient energy for heating, lighting and ventilation. In the case of existing educational premises, this purpose can be defined more specifically as to reduce space-heating loads while maintaining or enhancing the use of natural light and ventilation. In many overglazed schools of the 1960s, this can involve a reduction in glazing, even on a southerly aspect, to achieve an optimum balance. In a new building or major refurbishment, the glazing of courtyards or atria can provide a useful unheated space, and in this instance the purpose of passive designs is to maximise amenity for minimum energy use.

2.2 POTENTIAL IMPACT

Because of their form and use, new school buildings represent the most favourable category of non-domestic premises in which passive solar design techniques could be employed to pursue energy efficiency, as the second row to Table 2.1 illustrates.

However, because of the recession and the current decline in the number of pupils in schooling —the primary school population has fallen by about 20% since 1973 and the secondary is expected to drop by nearer 30% by 1991[1]—it is unlikely that a significant proportion of new buildings will be added to the country's stock of schools during the remainder of this century. In the short to medium term, this means that, if passive techniques are to be used to reduce fuel consumption in educational premises, they are more likely to make an impact if employed in retrofitting, remodelling or upgrading existing accommodation.

There is, at present, a lack of collected information on whether local authorities, who are responsible for erecting and maintaining the majority of Britain's educational premises, are employing passive techniques for this purpose. A preliminary, and largely theoretical, assessment of the possible contribution which the application of such techniques to these buildings could make to reducing fuel consumption has been attempted by Duncan and Hawkes.[2] This work is drawn on, enlarged and qualified here.

2.2.1 Passive potential of existing premises

The extent to which passive techniques can successfully be applied to existing educational premises depends upon a complex set of interrelated factors which includes:

(1) the pattern of fuel consumption in such buildings;

Table 2.1 Possible applications of passive design to non-domestic buildings[a]

Building type	Passive concept	Correct orientation/ shading	Passive		Passive or Hybrid (ie with or without fans)				Hybrid
			Non-diffusing—direct gain	Diffusing—direct gain	Atrium—direct/indirect/isolated gain	Conservatory—direct/indirect/isolated	Trombe wall—indirect gain	Thermosyphon with or without storage—isolated gain	Roof space collector with fan—isolated gain
Schools	Existing	000	00	0	00	000	0	00	0
	New	0000	000	0000	000	0000	00	000	000
Factories/ warehouses	Existing	00	0	0			0	00	
	New	000	0	00			0	0000	
Low rise offices	Existing	0	00	0	0	0	0	00	0
	New	000	00	000	0	0	00	000	00
High rise offices	Existing	00	0	00	0	0	00	00	
	New	000	00	000	0	0	00	000	000
Low rise flats	Existing	00	00	0	00	00	0	0	00
	New	000	000	000	0	0	00	000	000
High rise flats	Existing	00	00	0	0	0	0	0	
	New	000	000	000	00	0	0	000	
Low rise education residences	Existing	00	00	00	0	00	00	0	00
	New	000	000	000	0	00	00	000	000
High rise education residences	Existing	00	00	00	0	0	0	0	
	New	000	000	000	0	0	0	000	
Low rise nurses' residences	Existing	00	00	00	0	00	00	0	
	New	000	000	000	000	000	00	000	
Low rise hotels	Existing	0	00	00	0	00	00	0	00
	New	000	000	000	000	000	00	000	000
High rise hotels	Existing	00	00	00	00	0	0	0	
	New	000	000	000	0000	00	00	000	

[a]The number of zeros indicates the likely applicability of a passive measure based on physical constraints and building use:
0   1–20% applicability level
00   20–40% applicability level
000 40–60% applicability level
0000 60–80% applicability level.

From Duncan and Hawkes,[2] reproduced by permission of the Energy Technology Support Unit.

(2) physical characteristics of school buildings, e.g.
  — their orientation
  — their built form, i.e. plan form, section and construction;
(3) social, political and economic issues, e.g.
  — occupancy patterns
  — users' requirements and expectations
  — maintenance and replacement policies
  — funding and cost-effectiveness of techniques.

Only the first two sets of factors are considered here. In practice, they should all be viewed not just in isolation but jointly, in terms of their interaction and interdependency.

2.2.2 Fuel consumption

In governmental statistics on energy consumption, school buildings are often classed with offices and hospitals (see Fig. 2.1). This presentation is correct in that it does identify that space and water heating (combined) are the major source of fuel consumption in educational premises; but aggregation of schools with other types of building is misleading, because it conceals their comparatively insignificant consumption of electricity for artificial lighting (Table 2.2).

This comparative insignificance is, however, paradoxical. For it is now clear that significant fuel savings can come in school building from adopting passive means to reduce electricity consumption still further.[4] This can be achieved in new buildings by applying good daylighting practices during design. In both new and existing premises, it can result from installing more sophisticated manual or automatic switching controls in order to minimise the use of electricity when adequate daylight is available.

The particular pattern of energy consumption shown in Table 2.2 derives from a number of shared features, such as:

— the shallow plan forms of most school buildings;

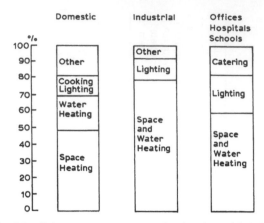

Fig. 2.1. Primary energy consumption by class of building. (Source: drawn by Cranfield Institute of Technology, figures taken from Energy Efficiency Office[3])

— the (until recently) mandatory provision of daylighting;
— their diurnal and seasonal patterns of occupancy.

It occurs despite reliance in British schools on natural ventilation (unlike many of their counterparts in, for instance, North America).

Collated national evidence that would enable energy consumption in primary and secondary schools to be disaggregated is unavailable. On the basis of the findings of their comparative analysis of the metered consumption of 116 secondary schools in Essex, Yannas and Wilkenfeld[5] noted that on average secondary schools used 4% less fuel per unit area than primary schools (see Table 2.3).

Table 2.3 Fuel consumption in Essex school buildings, 1974/75

	Average net fuel consumption per unit area (kWh/m^2)
Primary schools	279
Secondary schools	267

Source: Yannas and Wilkenfeld.[5]

Table 2.2 Energy used in educational premises, 1976

	Annual energy use (Petajoules)					Primary lighting energy as % of heating energy
	Space heating		Lighting			
	Delivered	Primary	Delivered	Primary		
Educational buildings	60.8	72.2	4.2	15.7		21.75
All non-domestic buildings		703.4		287.0		40.80

From Duncan and Hawkes,[2] reproduced by permission of the Energy Technology Support Unit.

The difference between the primary energy requirements of the two types of school is even larger. Those of secondary schools were 8% lower than the average for primary schools because of their lower use of electricity. Yannas and Wilkenfeld attributed the relative economy ·of secondary schools to a combination of factors including:

— higher gains from occupants due to differences of age and/or density of occupation;
— lower space temperatures;
— economies of scale (due, for instance, to multi-storey accommodation).

Other factors were differences in heating systems and seasonal efficiencies. Disaggregation of the end-uses of fuels in the sample of secondary schools is shown in Table 2.4.

Table 2.4 End-use of fuels in Essex secondary schools, 1974/75

	kWh/m^2	%	Fuel
Space and water heating	201	86.7	gas oil
Lighting and power	20	8.6	electricity
Cooking and kitchen hot water	11	4.7	natural gas

Source: Essex County Council.

2.2.2.1 *Fuel consumption: conclusions*

Generally stated, the purpose of a passive design is to ensure that the form, fabric and systems of a building are arranged and integrated in order to maximise the benefits of ambient energy for heating, lighting and ventilation. However, it should be evident from these analyses of fuel consumption that, in the case of existing educational premises, this purpose can be defined more specifically as to maintain or enhance the use of natural light and ventilation while reducing space heating loads.

2.2.3 Orientation

For the first half of the twentieth century, central government consistently and explicitly promoted school sites and buildings with southerly aspects (Fig. 2.2, for example). There is no documented evidence of the extent to which designers met this requirement. But it is reasonable to assume that most will have done so since, during this period, central government formally vetted local authorities' design proposals.

This suggests that, at least for buildings erected between 1904 and the 1950s, central government's recommendations will have enhanced solar access and aperture by encouraging good orientation and

Table 2.5 Age-related categorisation of school buildings

Age/plan type	Number of schools		Average floor area (m^2)	
	Primary	Secondary	Primary	Secondary
1. One or more large classrooms (mainly 19th century)	4 600 ⎤		300 ⎤	
2. Central hall or Ben Jonson (mainly pre-1919)	3 650 ⎦ 150		600 ⎦ 1 300	
3. Quadrangular (mainly 1919–1945)	550	300	1 500	6 000
4. Single corridor (mainly 1919–1945)	2 700	300	950	2 850
5. Finger plan (mainly immediately post-1945)	750	450	1 000	6 000
6. Post-1945 (other than finger plan)[a]	6 950	2 000	1 000	6 000
7. Combinations and unclassifiable[b]	4 100	1 850	850	5 700
Total schools	23 280	5 030		

[a]Including older schools brought up to regulation standards by remodelling.
[b]Including combinations of two or more of Types 1 to 6 or types unclassifiable in these terms.

From Duncan and Hawkes,[2] reproduced by permission of the Energy Technology Support Unit.

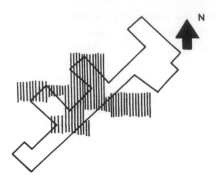

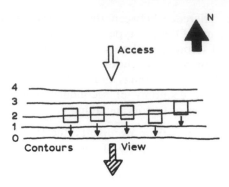

(a) On *flat sites* better and cheaper schools can be built if a compact design is substituted for a sprawling one.

(b) These site conditions permit the simplest design solution getting both sun and view.

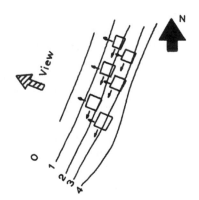

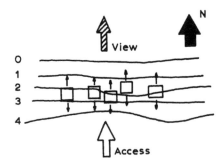

(c) and (d) Two types of site conditions where only by staggered plan arrangements can each class unit get both sun and view.

Note: Economy in siteworks can be achieved if buildings are always designed parallel with contours. To do this, and to achieve good view and sun conditions on varying types of sites will require a method of construction permitting flexible planning as (c) and (d).

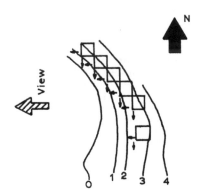

(e) This type of site condition calls for special planning arrangements to combine both sun and view to each class unit with economy of siteworks.

Fig. 2.2. Diagrams of the recommended relationship between buildings and site conditions. (From Ministry of Education[6])

discouraging overshading. On this basis, 4000 primary and 1050 secondary school premises (those with plan types 3, 4 and 5 listed in Table 2.5) are particularly promising candidates for passive retrofitting; but it is not these buildings that are likely to be selected by local authorities for upgrading, for reasons discussed below.

2.2.4 Physical characteristics of school buildings

2.2.4.1 Built form
School buildings can be characterised in terms of a series of distinct historical stages. These derive from changes in educational practices and/or the applica-

tion of scientific thinking and technological innovations to building design. So, in educational premises, there is a marked connection between the age of a building and its form—its internal spatial arrangement, section and construction. This connection was used by the Department of Education and Science to construct (simplified) age-related categories (see Table 2.5). Duncan and Hawkes[2] used these in their estimation of the passive solar potential of existing British primary and secondary school buildings (Tables 2.6 and 2.7).

The DES age-related categorisation can also be used to question which types of primary and secondary school buildings are the most attractive candidates for passive retrofitting by nature of their

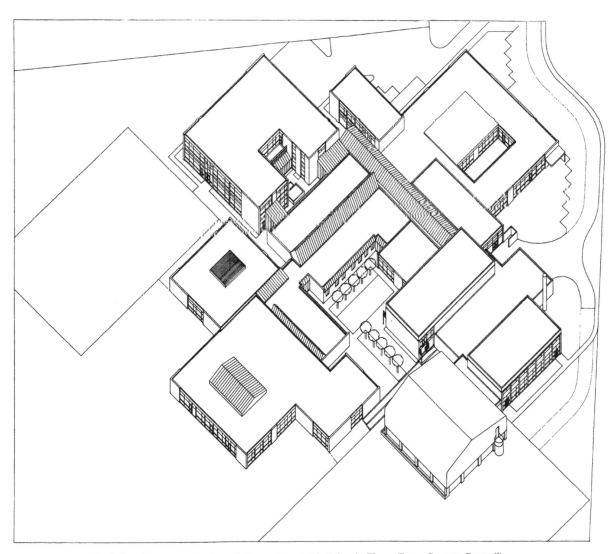

Fig. 2.3. Axonometric view of Thorpe Bay High School. (From Essex County Council)

Table 2.6 Estimate of the passive potential of existing primary school buildings

1		Typical building 2	3	4	5	Total UK stock 6	7	8	9	10	11	12	13	14	15	16	17	18	19	20	21
Building age/plan type or size group	Estimate	Floor area (m^2)	Useful solar (kWh/m^2)	Collector area (m^2)	Solar contributed (10^3 kWh)	Floor area ($10^6 m^2$)			Number of buildings (10^3)			Ultimate potential (10^6 kWh)		Fraction suitable		Technical potential (10^6 kWh)		Market penetration		Dateline contribution (10^6 kWh)	
						1979	2000	2020	1979	2000	2020	2000	2020	2000	2020	2000	2020	2000	2020	2000	2020
1. 19th century: one or more large classrooms	Low	300	60	70	4.2	1.4	1.3	1.2	4.6	4.3	4.0	18.1	16.8	0.3	0.3	5.4	5.0	0.05	0.25	0.3	1.3
	High		160	110	17.6							75.7	70.4			22.7	21.1			1.1	5.3
2. Pre-1919: central hall or Ben Jonson	Low	600	60	60	3.6	2.2	2.1	2.0	3.65	3.5	3.3	12.6	11.9	0.3	0.3	3.8	3.6	0.05	0.25	0.2	0.9
	High		160	100	16.0							56.0	52.8			16.8	15.8			0.8	4.0
3. 1919–1945: quadrangular	Low	500	60	100	6.0	0.8	0.8	0.8	0.55	0.55	0.55	3.3	3.3	0.8	0.8	2.7	2.7	0.05	0.25	0.2	0.6
	High		160	400	64.0							35.0	35.0			29.0	29.0			1.4	7.2
4. 1919–1945 single corridor	Low	950	60	100	6.0	2.6	2.6	2.6	2.7	2.7	2.7	16.2	16.2	0.8	0.8	12.9	12.9	0.1	0.4	1.3	5.1
	High		160	200	32.0							86.4	86.4			69.1	69.1			7.0	27.6
5. Post-1945: finger plan	Low	1 000	60	60	3.6	0.75	0.75	0.75	0.75	0.75	0.75	3.0	3.0	0.7	0.7	2.1	2.1	0.2	0.5	0.5	1.0
	High		140	140	22.4							16.1	16.1			11.3	11.3			2.3	5.2
6. Post 1945: traditional and SEAC, etc.	Low	1 000	60	40	2.4	9.5	9.5	9.5	6.95	6.95	6.95	16.6	16.6	0.7	0.7	11.7	11.7			2.4	5.9
	High		120	200	24.0							166.0	166.0			117.0	117.0			24.0	59.0
7. Combinations and unclassifiable	Low					3.45			4.1 (=20% of 1–6)					(+20% of 1–6)		7.5	7.4			1.0	2.9
	High															51.0	50.4			7.6	21.2
8. All schools	Low					20.7										38.8	38.8			6.0	17.8
	High															267.0	264.0			46.0	130.0

From Duncan and Hawkes,[2] reproduced by permission of the Energy Technology Support Unit.

Table 2.7 Estimate of the passive potential of existing secondary school buildings in England and Wales

1		2	3	4	5	6	7	8	9	10	11	12	13	14	15	16	17	18	19	20	21
	Typical building					Total UK stock															
	Estimate	Floor area (m²)	Useful solar (kWh/m²)	Collector area (m²)	Solar contributed (10³ kWh)	Floor area (10⁶ m²)			Number of buildings (10³)			Ultimate potential (10⁶ kWh)		Fraction suitable		Technical potential (10⁶ kWh)		Market penetration		Dateline contribution (10⁶ kWh)	
Building age/plan type or size group						1979	2000	2020	1979	2000	2020	2000	2020	2000	2020	2000	2020	2000	2020	2000	2020
1 & 2. Pre-1919: large classrooms and central hall	Low	1 300	60	140	8.4	0.20	0.2	0.19	0.15	0.15	0.13	1.3				Negligible					
	High		160	200	32.0							4.8									
3. 1919–1945: quadrangular	Low	5 000	60	400	24.0	1.8	1.8	1.8	0.3	0.3	0.3	7.2	7.2	0.8	0.3	5.8	5.8	0.05	0.25	0.4	1.6
	High		160	1 600	256.0							76.8	76.8			61.4	61.4			3.2	15.4
4. 1919–1945: single corridor	Low	2 850	60	300	18.0	0.86	0.86	0.86	0.3	0.3	0.3	5.4	5.4	0.8	0.8	4.4	4.4	0.1	0.4	0.5	1.8
	High		160	600	96.0							28.8	28.8			23.1	23.1			2.3	9.3
5. Post-1945: finger plan	Low	6 000	60	360	21.6	2.70	2.70	2.70	0.45	0.45	0.45	9.8	9.8	0.7	0.7	6.6	6.6	0.2	0.5	1.4	3.4
	High		160	840	134.4							60.4	60.4			42.2	42.2			8.4	21.2
6. Post-1945: traditional SEAC, etc.	Low	6 000	60	240	14.4	12.0	12.0	12.0	2.0	2.0	2.0	28.8	28.8	0.7	0.7	20.2	20.2	0.2	0.5	4.0	10.2
	High		160	1 200	144.0							288.0	288.0			202.0	202.0			40.4	101.0
7. Combinations and unclassifiable	Low					10.5			1.85 (=60% of 1-6)					(+60% of 1-6)		19.6	19.6			3.8	10.2
	High															174.2	174.2			32.6	88.1
8. All schools	Low					28.0										52.2	52.2			10.1	27.2
	High															464.5	464.5			86.9	235.0

From Duncan and Hawkes,[2] reproduced by permission of the Energy Technology Support Unit.

plan forms. Preliminary analysis of this kind has been done by Duncan and Hawkes. They made little discrimination between categories in terms of the possible range of 'useful solar gains'. Here their estimates are similar except for post-1945 primary schools (i.e. plan types 5 and 6) where they suggested that the top end of the range is lower (see column 4 of Tables 2.6 and 2.7).

They differentiated more strongly between categories in terms of potential 'solar collector area'. Here they judged inter-war (plan types 3 and 4) and post-1945 finger-plan (plan type 5) primary schools along with inter-war (plan types 3 and 4) and post-1945 finger-plan and traditional secondary schools (plan types 5 and 6) to deserve the highest rating.

Of these preferred types, quadrangle school buildings could well be retrofitted with atria (depending on the span of the quadrangle) or with conservatories, especially if the latter were used to enclose the open verandas which many of them possess. Finger-plan school buildings might also be retrofitted with conservatories, but these might not provide a buffer zone doubling as solar collection and circulation space, since existing access to classrooms in such premises is often on the northern side in order to preserve the southerly aspect. Whereas post-1945 secondary school buildings might offer neither of these two particular possibilities, the commonly-found dispersion of a variety of currently separate buildings around a shared site could be exploited to create buffer zones in their interstices, as explored by Essex County Council at Thorpe Bay High School (Fig. 2.3).

Consideration of construction reinforces the preference rank-ordering for retrofitting established above. It also reveals why, in practice, this order is unlikely to prevail. In 1942 Stillman[7] criticised inter-war school buildings as 'hopelessly out-of-date', not least because they were 'too solidly built'. In their place, he advocated prefabricated standardised buildings, of lightweight construction, which '... being easy to dismantle and of high salvage value, will not have to be retained when requirements change'.

After the Second World War, this approach to the design of school buildings became widespread. By 1970, consortia such as CLASP and SCOLA using prefabricated building systems accounted for over half, by value, of the total school building programme.[8] Such buildings tend to be of lightweight construction, with low levels of insulation, and highly glazed in order to meet the then required 2% daylight factor.[9] It is typically these buildings

that local authorities are currently having to retain and upgrade, despite Stillman's prognosis. This work is being undertaken not because of their poor environmental performance (high winter heat losses and summer overheating) but because their fabric now needs repairing or replacing.

2.2.4.2 Built form: conclusions

Inter-war quadrangle and post-1945 finger-plan school buildings are the most obvious and worthy candidates for passive retrofitting because their plan forms invite the addition of atria or conservatories and their heavyweight construction provides sufficient thermal mass to mediate against unacceptable temperature swings and summer overheating. Post-1945 lightweight and highly glazed buildings are less eligible because they do not fulfil either of these last two criteria. However, for maintenance reasons, it is the latter kind of buildings that local authorities are having to upgrade. Because of this, in the short term, priority may have to be given to establishing precisely what contribution passive measures can make to improving the environmental performance and fuel consumption of these more problematic premises.

2.3 TECHNICAL MEASURES AND POSSIBILITIES

In the widest sense, any building must be considered as a system; and since heat can be gained or lost through the walls and roof, any building can be considered as a passive solar system. Certainly, any building with a window can be considered as a passive solar system. The optimum use of windows in an integrated design is to maximise the benefits of ambient light and energy and to minimise the use of artificial heating and lighting. Such optimal design offers environmental advantages as well as economic savings. New building clearly offers the greatest scope for projects to develop and test various aspects of solar design; however, as new building will represent a very small proportion of the total for the foreseeable future, new building should be considered a test bed to establish the viability of various technical measures that can subsequently be applied to existing buildings.

2.3.1 New building

New school building provides an opportunity to evaluate and monitor the various passive solar techniques that are available. These are now considered, although, as already remarked, the greater potential remains in the existing building stock.

2.3.1.1 South-facing glass

The most obvious approach involves a high proportion of south-facing glass. A heavy construction is necessary to ensure that the intermittent gain provides a steady heating effect, avoiding large temperature swings, in both winter and summer. St George's County Secondary School (now St Mary's College), Wallasey, Cheshire (Fig. 2.4(a)), was built in 1961, and although well ahead of its time remains to this day one of the largest buildings in the world to be designed specifically for heating by passive solar, with 877 m² of south-facing glazing used to heat classrooms and ancillary teaching spaces.

A more recent example of such a construction is St Cleer County Primary School in Cornwall (Fig. 2.4(b)). The double-glazed classroom block faces due south and is lit by clerestory south glazing, giving sun penetration to the north side. The internal mass has been increased by 25% above that found in traditional construction by the use of dense concrete blockwork in both internal walls and a wall on the inside of the glazing. Such a device is known as a Trombe wall.

2.3.1.2 Trombe wall

Poulton Lancelyn Primary School in the Wirral (Fig. 2.4(e)) utilises a heavy wall directly behind double glazing. The wall absorbs and stores solar radiation, which can thus be used over a longer period. The traditional Trombe wall has slots top and bottom; these are closed off during the day, allowing the wall to absorb heat. When heating is required in occupied spaces the slots are opened and either natural or forced air flows up the warm wall surface and through slots into the occupied spaces. In the case of Poulton Lancelyn, the amount of heat is controlled by motorised dampers and variable speed fans.

2.3.1.3 Lightweight solar wall

One advantage of using passive solar in a school is that it requires intermittent heating, mostly during daylight hours. Extensive and costly heat storage is therefore not always necessary if a traditional means of heating is to be installed. In St John's School, Clacton, special external lightweight cladding was designed in the form of vertical ducts with glass external and steel internal. As with a Trombe wall, solar gain is collected in the space between the glass and inside wall; but in this case the

heat is not stored but passed directly to the inside when heating is required and vented to the outside otherwise. A thermostatically controlled damper passes the air in the appropriate direction.

2.3.1.4 Structural shading

Passive solar design in its most basic form is to utilise free heat and light when required and available, but reject the energy when it is not required and provide protection against overheating. Blinds, preferably external, provide one means of control; but they require constant human intervention. Structural projections above or to the sides of window openings, if carefully sized, can take advantage of the sun's varying altitude during the year to permit penetration only during the heating season and at morning and evening during other seasons. There are many examples of such design, one of which is Crossfields Schools in Berkshire (Fig. 2.4(f)), where cloisters provide summer shading to the ground floor and roof eaves to the first floor.

2.3.1.5 Atria

An atrium provides a most attractive increase in amenity for minimum energy use. In Nabbotts School, Chelmsford, Essex, (Fig. 2.4(g)), an atrium has been incorporated within the design cost limits and with no extra energy requirement. The area is in addition to teaching space and, although unheated, is used for cloakrooms and other facilities, allowing maximum use of heated teaching space. The atrium roof can be opened mechanically during warm weather.

2.3.2 Existing buildings: retrofit examples

Falling school rolls, of which education planning has had to take full account, mean that a significant amount of new building is unlikely to be started this century. If passive solar design is to make any impact on energy use it will be in the refurbishment of existing building stock. This is not at all as restrictive as it may at first appear, as in much of the existing building stock there is potential for use of passive solar.

2.3.2.1 1920–1950

Many schools built between the wars were of the quadrangle design, with large open courtyards and thus a very large external area. These buildings have a high fabric heat loss and, with doors and windows on two sides of classrooms, high infiltration rates. The construction of an atrium over the quadrangle

(a)

(b)

(c)

(d)

(e)

(f)

(g)

(h)

(i)

(j)

Fig. 2.4. (a) St Mary's College (previously St George's County Secondary School), Wallasey, Cheshire. Direct gain. (b) St Cleer County Primary School, Cornwall. Heavyweight direct gain. (c) Thorpe Bay High School, Essex. Atrium retrofit. (d) Barnes Farm School, Chelmsford, Essex. Covered atrium retrofit. (e) Poulton Lancelyn Primary School, Wirral. Trombe wall. (f) Crossfields Junior School, Berkshire. Solar shading. (g) Nabbotts County Junior School, Chelmsford, Essex. Atrium. (h) Netley County Infants School, Hampshire. Conservatory. (i) Gt Leighs County Primary School, Essex. Conservatory retrofit. (j) Ravenscroft County Primary School, Clacton, Essex. Direct gain collectors.

of such buildings reduces external area, reduces infiltration and benefits from solar gain. Such refurbishments are expensive, but feasible where a rationalisation of school building is necessary. Thorpe Bay High School (Fig. 2.4(c)) was one such, where, with the careful use of covered streets and courtyards, the capacity of the school was almost doubled with no increase in the existing heating plant capacity.

2.3.2.2 1950s
During the 1950s the post-war population bulge required a massive school building programme, and, not surprisingly, less attention was given to energy consumption or building life. Many of these buildings are of wooden studwork which is now in poor condition, requiring recladding or in many cases rebuilding. Recladding offers an excellent opportunity to improve insulation and, more pertinently, to optimise glazing areas and orientation to obtain maximum benefit from solar heat and daylight whilst avoiding the problems of summer overheating. Kingsmoor Junior School in Harlow is one of many schools in Essex which are being reclad with this design optimum in mind.

2.3.2.3 1960s
The latter end of the 1950s and the 1960s saw a move towards light system building and glass curtain walling. High daylight requirements and the availability of cheap energy encouraged designs which have left an uncomfortable legacy to today's school occupants. In winter, the large external glazed areas create a cold radiant environment and the low insulation causes high fuel bills. In summer, uncontrolled solar gain causes equally uncomfortable high temperatures. Fortunately the remedy is easier and less expensive. In Essex a patent glazing insulation system has been developed which can be installed from the outside with minimum disturbance to the occupants. It consists of rigid insulation bonded to steel sheet which is cut to existing window size and bonded directly to the glass. The external surface has a plastic coating and is both strong and durable, and maintenance free. One of the many schools which have benefited from this system is Black Notley High School. Calculated choice of panels to infill provides lower fuel consumption and higher radiant temperatures in winter; optimum use is made of passive solar, and maximum use of daylight and reduced summertime temperatures are ensured.

2.3.2.4 1970s
The relative discomfort experienced in lightweight buildings led in the 1970s to concrete construction of heavier weight and to reduced window areas. Insulation levels, especially in the early 1970s, were still relatively low. It seems that little benefit can be obtained from free solar gain in these constructions, although many would benefit from additional wall insulation. Some smaller schools were built with a small courtyard or quadrangle which provides the opportunity for the addition of a fixed or movable transparent roof to form a pleasant atrium. This provides extra amenity, reduces energy consumption and provides an excellent means of increasing the capacity of a school during an amalgamation or closure of a neighbouring establishment. Such a movable atrium has been installed in schools in Essex, including Barnes Farm School at Chelmsford (Fig. 2.4(d)).

Adding a conservatory to the south face of a heavy low insulated wall can also provide extra amenity as well as reduce overall energy consumption. The example shown here is Netley Infants School, Hampshire (Fig. 2.4(h)), which had conservatories in the original design. However, conservatories provide far greater potential in existing heavyweight building, although they are rarely justifiable in their own right and would no doubt only be incorporated during a major refurbishment.

2.3.3 Future technical developments

The two most significant items required for a successful solar design are optimised glazing and balanced thermal storage.

2.3.3.1 Glazing
The ideal glazing would have low thermal conductivity, controllable radiant heat transmissivity and high visible light transmissivity. Glass manufacturers are well aware of these requirements and for many years have conducted research to find the ideal glazing combination. In the 1960s very few special glasses were available; these concentrated on the reduction of heat gain, and unfortunately daylight was reduced disproportionately. Current developments provide a far more favourable balance between light and heat control, and many such solar control glasses are available for single-glazed use. Double-glazed options provide far higher insulating levels, but at present can only be considered viable if reglazing is necessary. For the future, research into durable low-emissivity

coatings for glazing may well provide the perfect cladding material.

2.3.3.2 Thermal storage

Research is underway on more concentrated forms of energy storage, especially the use of change of phase of various salts. Although these promise much for the future, cost limits will ensure that heavy conventional brick and concrete will remain the best storage media for the foreseeable future.

2.4 BARRIERS TO PASSIVE SOLAR DESIGN

Compared to other non-domestic building, there arc fewer technical barriers to the more widespread use of passive solar designs. The vast majority of educational building is owned, occupied and maintained by the local education authority, which is also responsible for the fuel bill throughout the life of the building. There is therefore no disincentive due to unknown resale value or to benefits that would go to the tenant rather than the owner. Educational building, though normally improving on building regulation requirements, is not subject to them and can therefore be designed according to criteria of energy use rather than fabric properties. However, there are barriers that apply for all building types.

2.4.1 Inadequacy of design tools

A passive solar design requires that designers take into account the balance between dynamic energy gains (equipment, solar energy and occupants) and energy losses.

Current computer models are unwieldy, and sufficiently accurate only in the hands of specialists. User-friendly design methods are inadequate at present when stressed, beyond their calibration limits, by passive solar designs.

Work on expert systems embracing a graduated set of design methods should be supported. Meanwhile, designers should look for paradigms to those designs which have been demonstrated, by good monitoring, to be successful.

Such a model has been commissioned by the Department of Education and Science from the School of Architecture, University of Wales Institute of Science and Technology, Cardiff. However, this model is a research tool and requires much further refinement before it could be used regularly as a design aid by busy architects and engineers. The UWIST project is to be published as a series of articles in the technical press.

2.4.2 Shortage of performance data

When designing a building, and before embarking on the detailed analysis phase, a conscious decision must be made to take full advantage of passive solar. In the case studies discussed here, the designers had generally broad experience of such design or had access to specialist analytical services. Normally this decision would be based on published design and performance data of which there is very little. A programme of monitoring is proposed by ETSU of existing passive designs and the results of this monitoring should be published as soon as possible.

2.4.3 Familiarity with design concepts

To bring about an increase in the awareness of design professionals (architects, engineers and builders) is a crucial first step in developing the capability throughout the building industry. In school building, an infrastructure for the dissemination of design information to the property design and service departments of local authorities could be created through existing LAMSAC and EEO activities. A joint ETSU/LAMSAC seminar specifically for passive solar design is recommended.

2.4.4 Costs and benefits

The lack of information on costs and potential benefits of passive solar design, especially in retrofit situations, deters most authorities from making any investment. Most measures are taken either as part of a larger basket of energy and envelope improvement measures or to provide a greater usable area for minimum energy cost. The financial benefits are poorly documented or unknown and the improvement in amenity is rarely considered.

A study should be implemented to detail both financial and amenity benefits. The latter is of considerable significance in educational building—in particular when, for instance, artificial lighting power requirements are reduced as a result of improved daylight.

The application of the design principles expressed here can usefully be studied by reference to their implementation in a county such as Essex; equally effective examples in other areas· could easily be cited, but schools in Essex that would provide interesting case studies are listed in Appendix 2A at the end of this section.

2.5 INSTITUTIONAL BARRIERS

2.5.1 Funding

2.5.1.1 Cost guidelines
Most passive design features in new buildings can be incorporated at no extra cost. Until the issue of an Administrative Memorandum by the Department of Education and Science (Elizabeth House, York Rd, London SE1 7PH) in January 1986, local authorities had to consider the cost guidelines set by the DES, but this is no longer true; so far as local education authority buildings are concerned, there are now no official cost guidelines that are applicable. However, most local authorities set a cost limit on their building projects, and items like special glazing systems are rarely justified on this basis. Hence local authorities should make additional allowance where increased energy savings justify it.

2.5.1.2 Central government funding
Before 1982 funds were made available to local authorities for energy conservation measures, and there is no doubt that many passive solar retrofit measures could meet justifiable payback criteria. However, this funding, which was intended to be 'pump priming', has not been available since. No further initiative is planned by DES. It is recommended that EEO consider giving priority to passive solar demonstration projects in schools and making grants to local authorities for passive solar retrofit measures as a national initiative.

2.5.1.3 Restrictions on local authority spending
Restrictions imposed by central government on the capital spending of local authorities are a severe disincentive to investment in any form of energy conservation. Those authorities that are able to allocate money have self-imposed criteria for investment levels and payback criteria which are formulated to ensure a financial return on capital investment. With such self-imposed criteria, it is recommended that central government discount energy conservation investment when setting spending limits.

2.5.2 Building architecture

2.5.2.1 Existing building
Their likely orientation alone makes existing school buildings the most attractive category of non-domestic premises for passive retrofitting. Throughout much of the twentieth century, design guidance has focused on the advantages of a southerly aspect for both school sites and buildings. In the late nineteenth century, Robson, who was consultant architect to the Education Department at Whitehall, only promoted the former:

> It is well known that the rays of the sun have a beneficial influence on the air of a room, tending to promote ventilation, and that they are to a young child very much what they are to a flower ... [but] the main lighting of the schoolroom should never be from the south or south-west, although sunny windows should be provided. The coolest, steadiest and the best light is that from the north and the principal aspect should first be selected as near that quarter as may be practicable ...

In 1904, the Board on Education[10] commended the value of a site 'open to the sun' for its effects not just on ventilation but on children's health as well. Attention to this last factor was increased by the 'open air' school movement, fuelled by misgivings about the health of recruits revealed by the Boer War. By 1919, it was argued by Clay[11] that

> ... the best aspect for the main lighting windows is south-east, which allows the sun to come in during the early hours, when it is most valuable, while ensuring that it is off the windows before the hot part of the day.

In 1933, when a committee of the RIBA[12] reported on 'the orientation of buildings', it placed almost as much importance on sunshine as on fresh air for schools. It recommended that

> ... school should be so placed on site that no shadow can fall on the buildings and every available ray of sunshine can play around and enter them.

The value of sunshine here was seen as stemming from its 'psychological effects' rather than its 'warming agency'. Three years later, the Board of Education[13] revised its own guidance and repeated the committee's recommendation almost verbatim. But it also acknowledged the beneficial contribution of the sun's 'radiant heat' to occupants' comfort in rooms facing in a southerly direction.

After the Second World War, when the Ministry of Education published substantially amended guidance,[6] it still endorsed the same criteria for choosing school sites and for positioning buildings on them:

> The most satisfactory [site] ... is a gentle slope down to the south or south-east, with contours running approximately east to west ... arrange the elements of the building, not only to suit the contours but to take advantage both of the sun and the view (if there is one).

As these selected statements reveal, for the first half of this century central government consistently and explicitly promoted school sites and buildings with southerly aspects.

It seems reasonable to conclude therefore that, until quite recently, school buildings have been well orientated for passive solar benefits.

2.5.2.2 New building

This assumption cannot be applied to very recent additions to the building stock. New school building must comply with DES Design Note 17, but no allowance is made in this design note for solar gain except for a requirement of 20% glazing with a view. In this respect, Design Note 17 gives less support to passive design for school buildings than the Building Regulations 1985 (in conjunction with the CIBSE Energy Code) for non-domestic buildings in general. Although DES accepts that passive solar should be considered, because of the variations in shading from other buildings in a town and the complexity of heating controls required to take full advantage of free gain, no advice is given. However, if design tools and techniques were readily available to support DN17 solar allowance could be made.

Development and research projects sponsored by DES form the basis for changes in guidelines. It is therefore recommended that joint EEO/DES projects be initiated to enable the benefits of passive solar to be incorporated in DN17 guidelines. It is further recommended that the advice on choice of sites and orientation of buildings which existed prior to 1949 be revised, updated and reinstated.

2.6 CONCLUSIONS

In conclusion, this sub-group considers that there is great potential for the use of passive solar energy in school buildings, particularly through the adoption of more sophisticated controls over electric lighting to increase exploitation of daylight. To realise this potential, technical barriers need to be overcome,

design tools and performance data need to be provided, familiarity with design concepts must be encouraged and more information must be provided on costing benefits.

Institutional barriers have to be lowered. Preferential funding facilities for energy investment and to encourage the adoption of passive solar may also have to be considered in design requirements and guidelines.

Despite these difficulties, much has been achieved, and, as this paper shows, many good examples now exist. The need is to use these successful examples to encourage others.

APPENDIX 2A: CASE STUDIES

The following schools provide interesting examples of passive solar design:

St Mary's College, Wallasey, Cheshire
St Cleer County Primary, Cornwall
St John's Primary, Clacton, Essex
Poulton Lancelyn Primary, Wirral
Nabbotts County Junior, Essex
Crossfields Junior, Berkshire
Netley County Infants, Hampshire
Thorpe Bay High, Essex
Black Notley High, Essex
Kingsmoor Junior, Essex

Some of these are among the schools shown in Fig. 2.4.

References

1. Department of Education and Science. *Annual Report 1980*. HMSO, London, 1981.
2. DUNCAN, I. P. & HAWKES, D. *Passive solar design in non-domestic buildings*. A report to the Energy Technology Support Unit, Ref. ETSU-S-1110, Martin Centre/Building Design Partnership, Harwell, June 1983.
3. Energy Efficiency Office. *A review: Energy Efficiency Demonstration Scheme*. Department of Energy, 1984.
4. CRISP, V.H.C, LITTLEFAIR, P., COOPER, I. & McKENNAN, G. *Daylighting as a passive solar energy option*. Final Report to the Energy Technology Support Unit, Building Research Establishment, Garston, Watford, 1986 (Contract no. ET 174/175/099).
5. YANNAS, S. & WILKENFELD, G. *Energy strategies for secondary schools in Essex: report on Phase 1*. Energy Programme Research Paper 1/78, Architectural Association Graduate School, London, 1978.
6. Ministry of Education. *New primary schools*. Building Bulletin, HMSO, London, 1949.
7. STILLMAN, C. Schools. *Architects' Journal*, **96** (1942) 342–51.
8. Department of Education and Science. *The consortia*. Building Bulletin 54, HMSO, London, 1976.
9. Department of Education and Science. *Lighting in schools*.

Building Bulletin 33, HMSO, London, 1967.

10. Board of Education. *Rules to be observed in planning and fitting up public elementary schools*. HMSO, London, 1904.

11. CLAY, F. *Modern School Buildings*. Batsford, London, 1919.

12. Royal Institute of British Architects. *The Orientation of Buildings*. RIBA, London, 1933.

13. Board of Education. *Suggestions for the planning of buildings for public elementary schools*. Education Pamphlet No. 107, HMSO, London, 1936.

Section 3

Use of Passive Solar Energy in Offices

John Campbell

Technical Director, Ove Arup Partnership, London

This paper presents the work of a sub-group, dealing with commercial buildings, of the Watt Committee Working Group on Passive Solar Building Design.

Membership of Sub-group

J. Campbell (Chairman)

Dr V. H. C. Crisp
D. M. Curtis
Dr D. Hawkes
G. K. Jackson
Dr D. Lindley
Dr J. G. Littler
Professor J. K. Page
P. H. Pitt

3.1 THE NEEDS OF THE OFFICE ENVIRONMENT

In the United Kingdom, most office accommodation is in buildings constructed and run for the purpose by privately owned organisations in commerce and industry, though there is also an important public sector. Generally, therefore, in office buildings, the value of using passive solar techniques is tested by its contribution to the profitability of the enterprise. By comparison, most schools (the subject of Section 2 of this Report) are owned and run by the local authorities and are largely financed from the proceeds of local and national taxation for purposes established by law.

3.1.1 Efficiency and comfort

Despite this institutional difference, the application of passive solar design has advantages in the office sector. They arise primarily from the needs of office workers for an environment that enables them to work efficiently in sufficient comfort. In the modern office, the interaction of office equipment with the environment is another important consideration. In the offices sub-group of the Watt Committee Working Group on Passive Solar Energy, it was necessary to begin by stating clearly the potential of passive solar energy as a contribution to the fitness of a building for its purpose. Accepting the views of the whole working group, expressed in Section 1 of the Report, the sub-group considered first the impact of the design approach to the utilisation of solar energy, and then reviewed the technical factors, noting the hurdles that the architect and builder must surmount in applying passive solar techniques in office buildings.

3.2 DESIGN APPROACH

Two principal types of buildings can be defined by the design approach. The two types are as follows:

(1) *Climatically interactive*: in which the building is designed to interact with its environment, in a way which is beneficial to both the building's occupants and its energy consumption;

(2) *Climatically rejecting*: in which an exclusive approach is used and the external environment is totally isolated from the internal environment.

There are occasions when the second of these approaches is the right one — computer rooms are an example — and it is possible to have one building in which some sections should be interactive and others exclusive. In some cases this can be turned to advantage by clever design, the waste from the exclusive portion being used to meet the requirements of the interactive portion. It should not be said, therefore, that it is wrong to use an exclusive approach; the right solution should be used in each case. It must be recognised, however, that when exclusion is necessary the opportunity to apply passive techniques is vastly reduced.

3.3 EFFECTS OF HEAT GAINS

Commercial buildings come in a wide range of types. At one extreme is the office full of dealers in the City, with internal heat gains as high as $140 \, W/m^2$. Alternatively, a traditional solicitor's office may have almost zero internal gain apart from the occupant himself. These two examples require vastly different solutions. In both cases there is a requirement for light, but whereas in one case heat is required, in the other it would be nothing but an embarrassment.

The reduction of energy consumption alone cannot be the only consideration. If energy consumption were the only criterion by which the success of a commercial building were judged, a windowless underground building would be the ultimate in successful energy-conscious design. Using high-efficiency lighting sources with modern application techniques, an underground building could be designed that would use no more than $80 \, kWh/m^2$ per year (about a quarter of a therm per sq. ft per year in Imperial units).

A typical value for an office can be seen in Fig. 3.1. The majority of the office space surveyed fell between 1 and 1.25 GJ per year[1] — about four times the figure mentioned for the underground building.

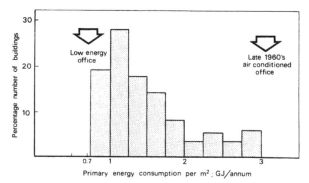

Fig. 3.1. Primary energy consumption in non-air-conditioned offices. (Courtesy Building Research Establishment, Garston.)

An underground building does not fully satisfy the requirements for a pleasant comfortable environment. Buildings without windows do not satisfy the need which humans still seem to have for light, view and general outside awareness. Since annual salary costs can be £1200/m², whereas the annual energy bill is unlikely to be more than £20/m², a drop in output of even a small amount cannot be considered.

This is why, for many years, energy was ignored in the operation of commercial buildings; then, when it was realised by building operators that any savings in this small figure travelled unchanged across the balance sheet and reappeared in the profit column, real progress was made.

3.4 ENERGY SOURCES

Energy in the United Kingdom is normally provided by gas-, oil- or coal-fired plant. Electricity may be added to this list — although the purists would disagree — because, as far as the building's owner is concerned, it is an energy source with a cost attached. Coal (or coke) was used primarily for heating until about 1955; after that it was steadily displaced by oil, which was not only cheaper, but could be provided with automatic controls. Gas started to make its presence felt in about 1970 (earlier in domestic use), as it had many of the advantages of oil and did not involve storage on site. Coal has recently started to make a comeback, which slowed down a little around 1984–85.

3.5 ENERGY USE

After the energy has been delivered to a building, it is used in one of several ways.

Lighting is always provided electrically, but different types of fittings have different luminous efficiencies (normally referred to as 'efficacy'). As an example, a tungsten bulb reproduces about 4% of its electrical input as light, whereas a fluorescent tube achieves about five times this value.[2]

Small power in commercial buildings now feeds typewriters, word processors and VDUs, but computer equipment has recently been subject to a definite change in progress. The growth in capacity of the microcomputer is resulting in distributed processing, with the effect that the loads on electricity supply now occur throughout the building and are not concentrated at one location. The VDU had a power requirement of around 50 W whereas the

modern micro which is displacing it requires about 120 W.

Lifts and/or escalators are essential for the efficient operation of high-rise buildings. From the point of view of internal heat gain, the lift motor is normally located away from the occupied space, whereas the escalator's motor is located in the occupied space.

The heating, ventilation and air-conditioning (HVAC) systems comprise boilers, refrigeration equipment, pumps, fans, cooling towers and all the associated control equipment.

How can the energy consumption of these items be reduced? In some cases it cannot. One thing that must always be remembered about a commercial building is that it is designed for a purpose: it is a place in which people work. If they are going to work happily and efficiently, the prime requisite is that the building should provide the right atmosphere. The small power requirements — the lifts, escalators and vending machines — are all required for efficient operation of the building. Lighting and thermal comfort are equally important.

3.6 ALTERNATIVE COURSES OF ACTION

In most cases a drop in standards is not acceptable, so the only ways to reduce energy consumption are to make the building interact with the external environment in a complementary manner and to redesign the energy-consuming items in the building so that they use less energy when performing the same task, or, better still, both.

3.6.1 Improved building design

Redesigning the equipment in the building is outside the scope of the building design team, although it is fair to say that there is considerably more feedback to manufacturers from designers than there used to be. Interacting with the environment, however, is totally the responsibility of the design team.

Historically, buildings have always been climatic moderators. It is only in the last 30 or 40 years that they have become enclosures for an artificial environment. Until recently, in terms of the length of time during which we have been constructing buildings, artificial lighting was provided by candles, oil and later gas; natural daylight was so markedly superior that buildings were always designed to take advantage of it. The Georgian façade of Southernhay West, Exeter (Fig. 3.2),

Fig. 3.2. Characteristic tall Georgian Windows at Southernhay West, Exeter. (Photograph by H. Sowden; architect Richard Ellis.)

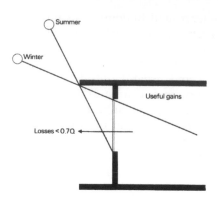

Fig. 3.4. Shading glazing. (Courtesy Ove Arup & Partners.)

shows the characteristic tall narrow windows which give a good compromise between light penetration and heat loss.

3.6.2 Daylight

Most modern commercial buildings have very high levels of internal heat gain, and heating (other than from cold after being empty) ceases to be a problem; so, whereas with some types of building there is a tendency to seek the benefit of both heat and light from the sun (Fig. 3.3), in the case of a commercial building the intention is to obtain daylight, create awareness of the exterior and minimise solar heat gain. As a result some form of shading becomes important[1] (Fig. 3.4).

Designing a building for maximum utilisation of daylight is not the entire solution. The wiring and the controls have to be arranged so that the operator of the building can take advantage of the daylight.[3]

3.6.3 Built form

Building form as a solution is very important, but unfortunately site constraints sometimes restrict the architect's choice of available forms (Fig. 3.5). This does not mean that low energy design is out of the question; it just means that the solution has to change with the problem. Recent building designs show that a pattern may be starting to emerge.

Where the site imposes little or no restriction on the design team, a long thin building with side daylighting is the obvious solution. When site restraints force a higher degree of site utilisation on the designers, however, different solutions have to be found.

If a low-rise solution will provide adequate floor area, light wells could be the answer. If a medium rise is essential, a covered atrium is now in vogue.[4,5]

Light-wells in buildings up to about 30 m high have been commonplace for many years, and not only for office buildings. Although the concept of the atrium dates from classical times, when it implied a central space open to the sky, it has come

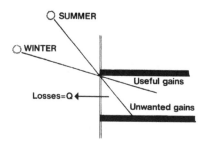

Fig. 3.3. Omission of shading. (Courtesy Ove Arup & Partners.)

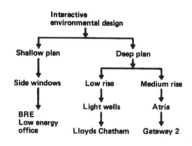

Fig. 3.5. Built form as a solution.

to be applied in a more sophisticated way in many modern buildings, and especially in offices. Essentially, the atrium is a volume of space in the middle of the building, constituting a considerable proportion — perhaps up to about 25% — of the whole; it provides a convenient way of heating, lighting and ventilating the surrounding parts, as well as perhaps serving social or circulatory functions for the occupants. It may be capable of admitting significant amounts of solar energy as a contribution to the operation of the building. Like other methods of harnessing solar power, the atrium nowadays may depend for its effectiveness on control systems of various kinds.[4]

Both of these solutions are methods of obtaining high site utilisation without the need to resort to deep-plan buildings. In each case it is possible to produce an example of the solution proposed.

3.6.4 Shallow-plan building

The shallow-plan solution shown in Fig. 3.6 is the low-energy building designed by the Building Research Establishment (BRE).[1] The daylight comes from the side windows and the solar gain is controlled by retractable shades. In winter this building uses a mechanical ventilation system; heat is recovered from the extracted air and warms the supply air. This alleviates the need to open windows for cooling when the external temperature is uncomfortably low. In summer, opening windows are used to control the temperature rise: the windows pivot, giving an opening at both top and bottom, to promote stack effect, and blinds are used to shade the glazing from direct radiation when there is a danger of overheating.

The internal environment of this building is very

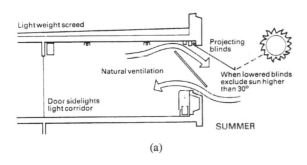

(a)

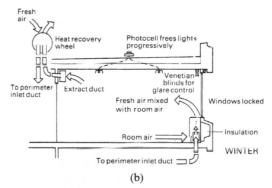

(b)

(c)

(d)

Fig. 3.6. BRE low-energy building. Typical south-facing office showing how comfort is achieved: (a) and (c) summer; (b) and (d) winter. (Courtesy Building Research Establishment, Garston; architects Property Services Agency.)

 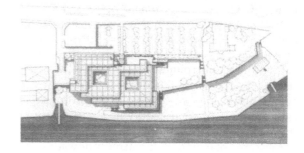

Fig. 3.7. Lloyd's building, Gun Wharf, Chatham, Kent. (Photograph by Crispin Boyle, courtesy Corporation of Lloyd's; architects Arup Associates).

pleasant, and the energy consumption compares very favourably with that of other building of the Property Services Agency, being about 60% (more nearly, 62%) of that of the main group of buildings shown in Fig. 3.1 and less than half of the average figure for all types of office building.

3.6.5 Light wells

There is a limit to the maximum depth that can be constructed in a building if natural ventilation is to be used. If the depth is too great, the required air velocities, particularly those near windows, become higher than is acceptable and the temperature distribution in the room becomes less even than is desirable. Following the deep plan route, the solution for low-rise buildings is to use light-wells. In the Lloyds building at Chatham, this solution was applied[6] (Fig. 3.7).

Most buildings have far more restrictions imposed on them than those of energy alone, but it is possible to take them into acount when balancing the other variables. The Lloyds building is next to a preserved building, the old rope works, in which ropes were made for the Battle of Trafalgar: not only was it not to be allowed to dominate, but also it had to tone in. This meant a high level of site utilisation. In fact, the courtyards are really atria in the later Roman tradition of courtyard gardens.

The result has been achieved without any penalty as far as the quality of the interior accommodation is concerned.

3.6.6. The atrium

The deep-plan medium-rise solution uses an atrium in the modern sense. The Wiggins Teape building at Basingstoke[4] is again designed without air-conditioning. In winter the atrium is used in a closed mode. Heat is provided by rejection from the computer air-conditioning system (Fig. 3.8). The summer operation uses the atrium to assist ventilation on the office floors.

The exterior looks quite normal, and again the working environment is very pleasant. The atrium itself provides a useful amenity and circulation space.

One of the main advantages of this type of building, with a covered atrium, is that light-coloured surfaces can be used without the danger that they may become dirty. The level of daylight achieved at low level is therefore much greater than would have been obtained with a traditional brick-faced light-well. This is not too important with low-rise buildings, but it is very important in medium-rise buildings where the view of the sky is restricted.

The technique is not restricted to new buildings. It can be applied to a refurbishment project (Fig. 3.9). What tools are necessary to design buildings like these? The summertime temperature calculation procedure devised by the Building Research Establishment has done Trojan service in this field,[7] and developments in computer capacity now enable more fundamental approaches to be adopted.

SUMMER

WINTER

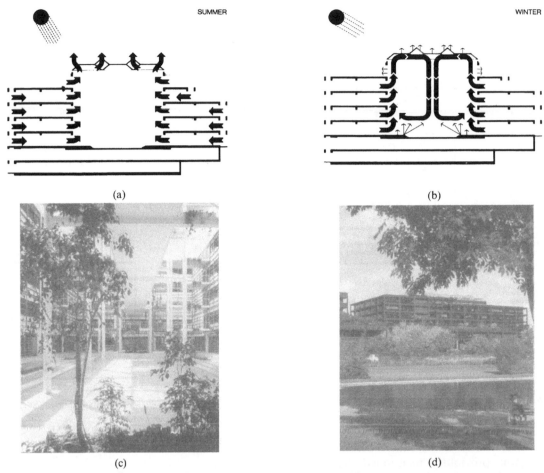

(a)

(b)

(c)

(d)

Fig. 3.8. Wiggins Teape (UK) plc, Gateway House, Basingstoke, Hampshire. (Photographs by P. Cooke, architects Arup Associates; Fig. 3.8 (d) courtesy Wiggins Teape (UK) plc.)

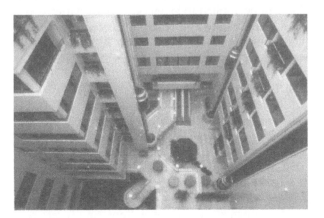

Fig. 3.9. Refurbishment of Royal London House. (Photographs by H. Sowden, architect Shepherd Robson.)

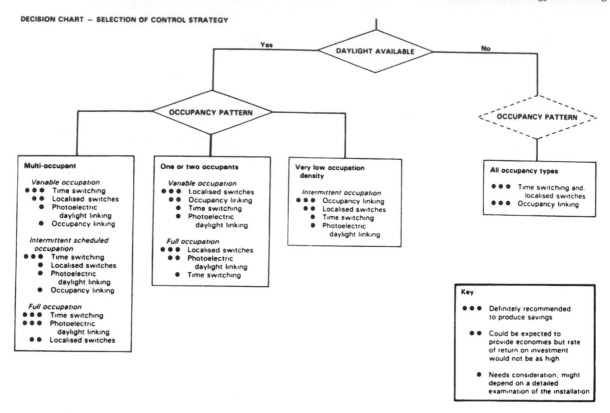

DECISION CHART — SELECTION OF CONTROL STRATEGY

Fig. 3.10. Decision chart—selection of control strategy. (Courtesy Building Research Establishment, Garston.)

3.6.7 Lighting control

Light, as was previously mentioned, has to be suitably controlled if benefit is to be obtained from daylight design. Far more detail is given in Fig. 3.10 than would normally be shown, but simplification would render it meaningless. The different strategies that can be applied to given occupancy patterns are shown and ranked in order of probable payback. This illustration is taken from *BRE Digest 272* which contains general advice on this topic.[8]

3.6.8 Infiltration

Infiltration models which have been produced initially for other purposes[9] can be developed as in Fig. 3.11. This model can cope with atrium design. As is well known, it is all too easy, in a tall building, for a window opened on the ground floor to have the desired effect of letting in cool air, whereas a window opened on an upper floor simply lets out hot air that has risen from a power source at a lower level in the building.

The program shown in Fig. 3.12 gives some idea of the area ratios that should be used to give reason-

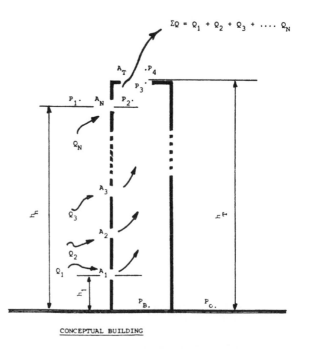

Fig. 3.11. Diagrammatic section through atrium: model development by M. Holmes. (Courtesy Arup R & D.)

```
            NATURAL DEMONSTRATION RUN    (iv)
    ----------------------------------------------------------------

    PREDICTED AREAS FOR SELF BALANCING NATURALLY VENTILATED BUILDING
    ----------------------------------------------------------------

        EXTERNAL  TEMPERATURE  25.00  (C)
        INTERNAL  TEMPERATURE  27.00  (C)
        HEIGHT OF TOP OPENING  15.00  (M)

    FLOOR VOLUME AIRCHANGE REF.HEIGHT    FLOOR OPENING FOR GIVEN TOP OPENINGS
          (CU.M)   PER HOUR    (M)     38.4   41.9   52.3   69.8  104.6  174.

        1  2000.0    5.00      0.50     4.8    4.8    4.7    4.6    4.5    4.
        2  2000.0    5.00      3.50     5.6    5.5    5.3    5.2    5.1    5.
        3  1750.0    6.00      6.50     7.1    6.9    6.6    6.4    6.3    6.
        4  1500.0    6.00      9.50     8.3    7.9    7.3    7.0    6.7    6.
        5   750.0    8.00     12.50    15.6   11.7    8.7    7.5    6.9    6.
    ----------------------------------------------------------------

    NOTES :
    IT IS ASSUMED THAT THE OPENING TO OUTSIDE IS THE OF SAME AREA AS THAT TO
    THE COMMON INTERNAL SPACE.
    EACH OF THESE IS EQUAL TO THAT GIVEN IN THE ABOVE TABLE (SQ.M)
```

Fig. 3.12. Computer output showing predicted areas for self-balancing naturally ventilated building. (Courtesy Arup R & D.)

able control of natural ventilation and ensure that the incoming air all travels via the route that was intended by the designer.

3.6.9 Thermal comfort

In assessing the level of comfort provided, Fanger's work[10] cannot be ignored. The area in which office workers can adjust their own comfort by varying their clothing level within normally acceptable limits is shown in Fig. 3.13.

3.7 ENERGY ANALYSIS

There are energy analysis programs that take a building, simulate its response to weather data and then examine the way in which the systems in the building react to the loads that would result and calculate the energy consumption of their component parts, such as fans, pumps, boilers and refrigeration equipment. Other programs, though not quite as comprehensive, simulate the load; this is useful in assessing the performance of the building

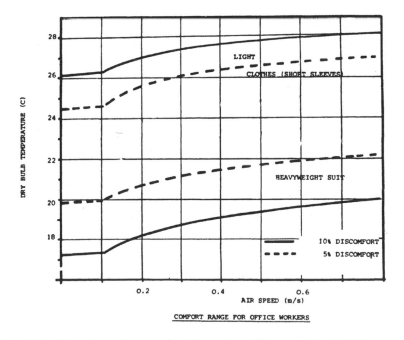

Fig. 3.13. Fanger's comfort zone for office workers. (Courtesy Arup R & D.)

envelope. Some problems are well taken care of, but others still require attention.[11,12]

Analysis programs are all very well, but, as the name implies, a very advanced design is needed before they can be applied. More design tools that would enable a designer to work for the correct solution from the start, when the question has not yet been confused by special factors, are necessary. Simulation can then be applied strictly for fine tuning. There are also design problems associated with fire precautions. Not enough is known about the behaviour of fires in atria, and until it is understood a degree of caution must be exercised. Perhaps restrictions should be based on fire load rather than volume.

The designer should remember that nothing is static. Other buildings will be constructed around his new creation—possibly on the south side, shading it from solar gain; possibly on the north side, with reflective glass that causes unwanted gain. Current planning and building control procedures do not protect the owner of a building against these factors.

3.8 CONCLUSIONS

One thing must be borne in mind, however. Although codes of practice have been introduced which should preclude bad buildings, guides and codes of practice do not, if followed in isolation, necessarily produce a good building. Besides, good buildings can be produced without codes. The buildings that were described earlier predate the current codes. The architect's obligation is to produce a good building that complies with the codes.

What would be the overall effect on the energy consumption of the country at large if all buildings were designed to the same standard as the buildings illustrated here? The total office floor area in the UK has been estimated at about $120\,km^2$, consuming some 10.7 mtce per year. As was shown in Fig. 3.1, energy consumption of the typical office is of the order of $1–1.25\,GJ/m^2$ per year. In some offices, the figures rise to as much as $3\,GJ/m^2$ per year,[13] but there could well be special equipment in these cases—as in computer rooms or canteens—so they are ignored here from the point of view of compatibility. This figure also compares quite well with the figure of 1 therm per square foot which was regarded as satisfactory in a good building in the late 1960s and early 1970s. The BRE building uses

only about 60% of this; if all building were constructed to this standard, a saving of about 4.3 mtce per year could result.

Looked at from a different point of view, on a national basis, the office stock in the UK is increasing by about 3% per year; this is comprised of some 3.6% new-build and 0.6% demolition. If this rate continues, by the year 2000 office floor area will have increased by 60%, and 45% of the total will be of modern design. If this new-build uses energy-conscious design principles, a 60% increase in office area will have been achieved with only a 33% increase in total energy consumption.

If all the existing stock were refurbished at the same time, it would be possible, at least in theory, to have a 60% increase in office area with no increase in energy consumption.

One question that is probably already emerging concerns the acceptability of this type of building to developers. It is true that all three of the buildings that are used as examples here were designed for specific clients and were in no way speculative. This is also true of other well-known buildings in this category, such as the British Telecom office. This, though, is probably the way in which new development must always start. Developers are in the market to provide the customer with the commodity that he wants, and there are strong signs that the occupiers of buildings now prefer to provide a slightly higher standard of accommodation for their staff.

The refurbishment project shown in Fig. 3.9 was speculative,[14] as also is the building at No. 1 Finsbury Avenue, in the City of London, shown in Fig. 3.14, which has a beehive-shaped atrium.[15] The floors, in practice, are shallow in plan, as no-one is far from a window—either an external one or one that overlooks the atrium. This type of building is really a hybrid. It is in a busy city-centre location, and although, for this reason, it has to be air-conditioned, it still incorporates some of the more attractive passive features which, in turn, contribute to reducing its energy consumption.

Good design, therefore, is a compromise between several different factors. They are:

(1) suitability for purpose;
(2) construction cost;
(3) flexibility;
(4) energy efficiency.

Energy efficiency should not be understated. Although there is a limit to the cost savings that can

It could well be that present concerns about energy will prove to be a blessing in disguise and will impose a style on our buildings as definite as those of previous ages. Perhaps there are signs of this already.

Fig. 3.14. Beehive-shaped atrium. (Photograph by H. Sowden, architects Arup Associates, client Rosehaugh Greycoat Estates Ltd.)

be obtained from lighting, it should be remembered that the maximum availability of daylight occurs at the same time of year as maximum solar gain. Careful design can mean that, at best (in the British climate), the need for air-conditioning may be eliminated and, at worst, not only may the running cost of the air-conditioning plant be reduced, but also, by reducing the size of the plant, its capital cost may be minimised also.

REFERENCES

1. CRISP, V. H. C., FISK, D. J. & SALVIDGE, A. C. *The BRE low energy office.* ECDPS Report. Building Research Establishment, Garston, Watford, 1984.
2. *CIBSE Interior Lighting Code.* CIBSE, London, 1983.
3. CRISP, V. H. C., LITTLEFAIR, P., COOPER, I., & McKENNAN, G. *Daylighting as a passive solar energy option.* Final Report to the Energy Technology Support Unit, Building Research Establishment, Garston, Watford, 1986 (Contract no. ET 174/175/099).
4. Gateway 2. *Architects' Journal,* **180** (46) (1984) 55–66.
5. SAXON, R. *Atrium Buildings: Development and Design.* Architectural Press, London, 1983.
6. Chatham. *Architects' Journal,* **173** (5) (1981) 199–217.
7. *CIBSE Guide Section A8.* CIBSE, London, 1986.
8. BRE Digest 272. *Lighting controls and daylight use.* Building Research Establishment, 1983.
9. International Energy Agency. *The validation and comparison of mathematical models of air infiltration.* Technical Note AIC 11, 1983.
10. FANGER, P. O. *Thermal Comfort, Analysis and Applications in Environmental Engineering.* Danish Tech. Press, Copenhagen, 1970.
11. BLOOMFIELD, D. P. Influence of the user on the results obtained from thermal simulation programs, *Proc. 5th Int. Symp. on the Use of Computers for Environmental Engineering Related to Buildings,* Bath, 1986.
12. International Energy Agency. *Buildings and Community Systems—Annexe 1 Final Report.* 1981.
13. KASABOV, G. (ed.). *Buildings—The Key to Energy Conservation.* RIBA, London, 1979.
14. Triton Court. *Architects' Journal,* **181** (7) (1985) 41–51.
15. Finsbury Avenue. *Architects' Journal,* **178** (34, 35) (1983) 65–7.

Section 4

Industrial, Retail and Service Buildings —Options for Passive Solar Design

David Lush

Technical Director, Ove Arup Partnership, London

&

Jim Meikle

Partner, Davis, Langdon & Everest (formerly Davis, Belfield & Everest), London

This paper presents the work of a sub-group, dealing with industrial, retail and service (IRS) buildings, of the Watt Committee Working Group on Passive Solar Building Design. As stated on page 59, at the Consultative Council Meeting the views of the sub-group were presented by Mr W. B. Pascall.

Membership of Sub-group

W. B. Pascall (Chairman)

Dr D. M. L. Bartholomew
M. E. Finbow
R. Hitchin
G. K. Jackson
D. M. Lush
J. L. Meikle

4.1 DIVERSITY OF BUILDING TYPES

4.4.1 General

The objective of this Section of the Report is to assess the passive solar design potential in a wide sector including industrial, retail and service buildings (IRS) and to indicate the possible energy savings in such buildings by implementing suitable measures over a period of time. Where possible, the institutional actions that would achieve the market potential are identified.

The IRS sector differs from the commercial, domestic and educational sectors because the building types and usages vary widely and there is a paucity of reliable stock and energy statistics. Table 4.1 summarises building categories included in the sector and provides indications of the current total stock and its potential for energy savings from passive solar design measures. Actual estimated potential savings for industrial buildings and hospitals are indicated in Appendix 4A.

Figure 4.1 illustrates diagrammatically how examination of building parameters, building stock and energy consumption for a particular building category can be combined to yield an indication of passive solar potential for that category. This potential can then be modified for external constraints to indicate possible energy savings. Application of this approach is discussed in Section 4.2 and illustrated in Appendix 4A.

4.1.2 Methodology

The approach adopted has been to develop a methodology that may be applied to all categories within the sector, as far as existing data permit. The methodology is also suitable for use on other sectors. Some categories have been fully treated in accordance with the methodology to validate its application. The aim has been to identify what passive solar design features can provide, in terms of energy savings on existing stocks of each category, and then to extend the approach to the building stock up to the year 2015. The overall exercise is intended to characterise building categories which can range from low-quantity (building area) high passive solar potential to high-quantity low passive solar potential.

4.2 POTENTIAL IMPACT

4.2.1 General

Studies in the IRS sector (and in any other building sector) should be geared to those buildings that show the greatest potential savings from passive solar design. The priority for studies needs to be gauged against a number of identified elements which affect this potential. In the first instance, the ranking will be based on a collective evaluation of the elements. The second stage is to identify (a) which of the building categories will offer the greatest potential in terms of replication, by size of floor area for the building types; (b) the energy saving from such replication; and (c) the cost-effectiveness of any passive solar design measures for the category in question.

4.2.2 Methodology

The assessment procedure, or methodology, illustrated in the Appendix 4A is set out below.

For the IRS (or any) sector:

— identify the parameters that most affect the effec-

Table 4.1 IRS building categories and stock statistics

Category	Existing stock floor area $(m^2 \times 10^6)^a$	Potential saving from passive solar design (savings code)[b]	
		Thermal	Lighting
Industry	226	1	2
Warehousing	128	2	2
Commercial stores (department stores)	11	2	1
Retail shops (inc. banks and restaurants)	76	1	3
Hospitals	32	3	3
Other health buildings	6	3	3
Residential institutions (inc. prisons)	14	3	3
Hostels (inc. nurses and students)	13	2	2
Hotels	25	3	2
Leisure/recreation/ conference	20	3	3
Sports	6	3	4
Agriculture	n.k.[c]	4	1
Public services/ utilities	n.k.	n.k.	n.k.
Religious	n.k.	n.k.	n.k.

[a] Figures for 1982–83.
[b] Savings code:

1	$\leqslant 1\%$
2	$> 1 \leqslant 2\%$
3	$> 2 \leqslant 4\%$
4	$> 4 \leqslant 8\%$
5	$> 8\%$.

[c] n.k. = not known.

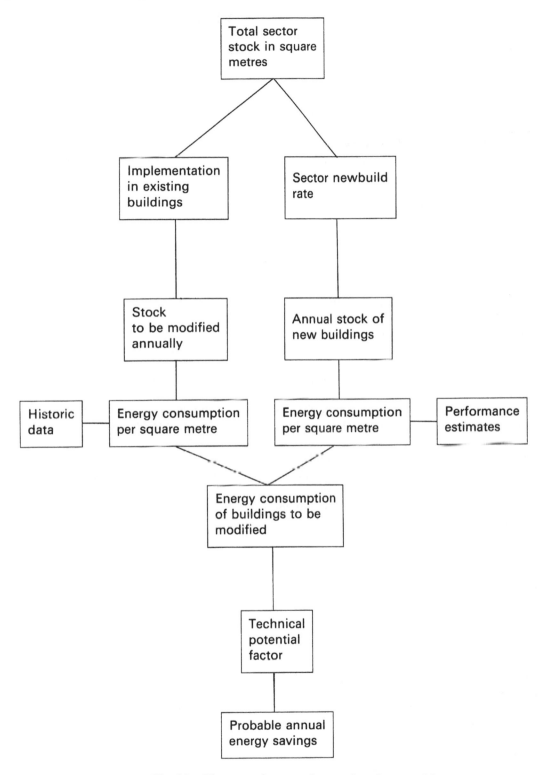

Fig. 4.1. Diagrammatic approach to passive solar potential.

tiveness of passive solar design;
— classify buildings within the sector by category, e.g. Industrial.

For each category:

— schedule the various combinations of parameters in most common use, as types, e.g. 1, 2, *et seq.,* within the selected category;
— classify the characteristics of each building type under each parameter, e.g. continuous, intermittent and variable usage;
— rank the passive solar potential on a scale of 0–10 for both thermal and lighting aspects (zero being a 'non-starter', 10 being ideal);
— from available national statistics and estimates, indicate the total category area and annual energy consumption and allocate proportions by floor area of the stock to each building type;
— for each building type combine the proportion of floor area with the thermal and lighting rankings to produce weighted rankings;
— sum these weighted rankings to produce aggregate passive solar potential rankings for the category;
— indicate the potential energy savings for all existing building types, using the weighted rankings;
— estimate the annual percentage new build rate for the category in question, together with the estimated percentage stock loss each year;
— estimate the potential savings from the annual new build, based on the weighted rankings and base energy consumptions similar to those for existing buildings;
— extend this to a graphical representation for potential savings 15 and 25 years hence, for predicted new build and demolition for each category;
— identify the most relevant passive solar measures applicable to those categories that qualify for follow-up and evaluate their validity against the estimates of energy saving and in terms of their cost-effectiveness—a very simple example is given at the end of Appendix 4A;
— possibly, where stock and performance figures are available in a suitably disaggregated manner, deal with each building type separately. It would then be possible totally to exclude particular types with a passive solar potential below a selected threshold. In the absence of this type of data, estimates of energy savings at category level have been made.

4.3 IMPLEMENTATION

4.3.1 General

The implementation of passive solar design in new and existing buildings involves building form, orientation, materials and services. In the case of existing buildings it is difficult to alter form and orientation, but renovation often provides the opportunity for modification of materials and services as an integrated system. In new buildings, consideration should be given to the important implications of orientation and form.

The methodology applied in the IRS sector is equally relevant to other sectors.

4.3.2 Existing buildings

The main options for passive solar design in existing buildings are:

— to make alterations to the fabric of the building, e.g. replace glazing with combinations of more effective types of glass and insulated, or blank, panels;
— to add conservatories and atria;
— to optimise space utilisation to suit orientation and function;
— to install control systems, including (a) time controls linked to occupancy patterns; and (b) thermostatic controls to improve the utilisation of solar gains.

4.3.3 New buildings

The potential for passive solar design in new buildings is greater than in existing buildings. However, it is dependent on a multi-disciplinary design approach which integrates building orientation, areas and orientation of glazing, insulation, building plan, spatial layout and building services selection, zoning, distribution and control. There is a need for building designers to consider carefully the practical options available to them. Good design should integrate the effective use of passive solar elements.

It is important to note the diversity of the buildings and even parts of buildings in the IRS sector. Commercial greenhouses would not operate without solar gain. Hospitals could, but wards, as opposed to operating theatres, would benefit through the sensible use of daylight and useful solar gain.

There is a need for a simple and robust approach to cost-effectiveness. Such an approach has been developed for housing and is described in Section 5, while the example in Appendix 4A also uses a simple evaluation method.

Passive solar design is unlikely to result in widespread changes in the style of IRS buildings. It could have an influence on the form and external treatment of the building. However, there are opportunities for better use of glazing materials, conservatories and atria. The use of photo-electric and thermostatic controls would allow the integration of glazing, lighting, heating and the needs of the occupants.

4.4 CONSTRAINTS AND OPPORTUNITIES FOR PASSIVE SOLAR DESIGN

4.4.1 General

There are three main issues to be addressed in considering the potential of passive solar design for IRS buildings: regulations, building providers and consumer demand. There is no available evidence to suggest that mandatory or advisory regulations act as a severe constraint on passive solar design. On the other hand, explicit recognition of the benefits of passive solar design could have a significant impact.

The resistance of building providers is perhaps more imaginary than real. Positive passive solar design features need not increase the initial capital cost of a project. Given suitable design guidance the resistance of designers and developers might be overcome. The guidance needs not only to cover the technical solutions, but also to highlight the fact that, for little or no additional capital costs, improved environmental quality and decreased running costs can be achieved. It may be possible to transform these improvements into increased rental income.

Consumer demand could be stimulated by lower bills for the use of utilities in passive solar buildings. Consumer ignorance of the benefits of passive solar buildings is a barrier to this demand. Professional and national media coverage will be essential if the necessary education and consumer awareness are to be successfully promoted.

4.5 CONCLUSIONS

In working on this section of the Report, a major conclusion has been that, in the IRS sector, there is a clear need for more comprehensive data, to a standardised format, on building stock and energy usage. The availability of such information would assist developers, designers and users to become more aware of the realities of energy-efficient buildings and enable more realistic energy targeting and monitoring.

Unlike the buildings dealt with in other sections of the Report, the available information on IRS buildings is insufficient, and therefore it cannot be said with certainty whether the potential gains are realised in practice.

We believe that there is a substantial potential to be demonstrated within the IRS sector for the reduction of energy consumption by the application of passive solar design in exisiting buildings and in new buildings. Buildings in some categories have a relatively low potential individually, but the stock is very substantial so a significant aggregate saving might be made. The savings could not be equally distributed across all building types; some are potentially better than others, and there is a need to identify the most promising.

In the short- and medium-term future, the greatest total saving can be achieved in the existing stock of buildings. The key opportunity point for these savings will occur when such buildings are refurbished over the next 20 years. In new buildings, if every opportunity is taken, passive solar gain could be optimised in the annual increase of 1% to 5% in the building stock.

BIBLIOGRAPHY

Davis, Belfield & Everest. Unpublished survey of national stock figures for sports halls, squash courts and swimming pools, from Sports Council Data, August 1985.

Department of the Environment. *Commercial and Industrial Floorspace Statistics. England. 1981–1984.* Statistics for Town and Country Planning, Series 11, Floorspace, No. 13. HMSO, London, 1985.

DUNCAN, I. P. & HAWKES, D. *Passive solar design in non-domestic buildings.* A report to the Energy Technology Support Unit, Harwell, Ref. ETSU-S-1110, Martin Centre/ Building Design Partnership, June 1983.

ETB Registration Unit. *Regional known stock of accommodation and bedspace capacity.* January 1983.

FITZJOHN, M. *National Swimming Pools Study: Swimming in the Community.* North Western Sports Council, 1983.

LAWSON, F. R. *Energy Use in the Hotel Industry of Great Britain.* Energy Conservation Demonstration Projects Scheme, Energy Technology Support Unit, Harwell, October 1983.

OLIVIER, D., MIALL, H. NECTOUX, F. & OPPERMAN, N., *Energy efficient futures: opening the solar option.* Earth Resources Research, London, 1983.

STAPLETON, M. Hotel and holiday accommodation — operators and consultants. AJ Update. *Architects' Journal*, **178** (27) (6 July 1983) 63–4.

APPENDIX 4A

INDUSTRIAL, RETAIL AND SERVICE BUILDINGS (IRS) SECTOR: ASSESSMENT OF PASSIVE SOLAR POTENTIAL

The assessment procedure for IRS buildings is based on the use of a standard matrix format shown in graphical form in Fig. 4.1 and with the headings indicated in Table 4A.1. This identifies the building categories in the sector; the total area of existing stock and total current annual energy consumption; and the major parameters that affect and influence the potential for passive solar design and savings. Each category is then divided into types, identified by a combination of sub-divisions under each of the major parameters. Each combination is then ranked for passive solar thermal and lighting potential on a scale of 0–10 (zero being a 'non-starter' and 10 being ideal). Each building type is then allocated a proportion of the total area stock of this category, and weighted rankings for each of the thermal and lighting potentials are then calculated.

All the rankings in these examples have been made by one individual, and although it is acknowledged that they are by no means perfect, any intuitive input tends at least to be consistent. It is reasonable to assume, for example, that all the rankings of 3 are consistently high or low, rather than that some should be 1 or 2 and others 4 or 5. Other assessors might wish to alter these rankings.

An example of the assessment method (or difficulty) concerns internal gains; these are one of the major parameters. A particular fabric combination with a low internal gain may have a high ranking, whereas the same building with high internal gains may be ranked much lower.

Completed formats for industrial buildings and hospitals are shown in Tables 4A.2 and 4A.3 respectively. Taking the industrial buildings as an example, both for simplicity and because the estimate is believed to be realistic, the aggregate rankings are taken to be equivalent to percentages of space heating and lighting demand. On this basis, for the current stock, approximately 0.8% (mid-point of 0.62 and 0.95) of total space-heating demand and 1.25% (mid-point of 1.03 and 1.46) of lighting demand could be saved. This is unlikely to be achieved in the existing stock for the following reasons:

(1) A retrofit programme would have to be implemented over a period of years; during that time, stock is being depleted or getting older.
(2) For one reason or another, not all buildings would be modified to attain their passive potential.

The total current stock of industrial buildings is approximately $225\,000\,000\,\text{m}^2$, and it is estimated to be growing at around 0.5% per year. This growth rate comprises average losses of 1% per year and average gains of 1.5% per year; Figure 4A.1 illustrates this, and all the new stock could similarly be designed to include passive solar features.

It seems likely that no more than, say, 5% of the total stock (more than $11\,000\,000\,\text{m}^2$) could be modified annually until a maximum of perhaps 65% of the current stock was reached in, say, 1997/98 (by which time it would represent almost 80% of the then current stock). This possible penetration of passive solar modifications is shown cross-hatched in black in Fig. 4A.1.

Demand for space-heating in the current stock is at an average rate of $2.10\,\text{GJ/m}^2$ and demand for lighting is $0.37\,\text{GJ/m}^2$. The annual saving in space-heating energy in 1985 arising from passive modifications could, therefore, be $0.189\,\text{PJ}$ ($225\,000\,000\,\text{m}^2 \times 2.10\,\text{GJ/m}^2 \times 0.8\% \times 5\%$), building up over 13 years to $2.457\,\text{PJ}$ per year in 1997. The annual saving in lighting energy demand in 1985 from passive modifications could be $0.052\,\text{PJ}$ ($225\,000\,000\,\text{m}^2 \times 0.37\,\text{GJ/m}^2 \times 1.25\% \times 5\%$), building up over 13 years to $0.676\,\text{PJ}$ per year in 1997.

As regards newly-built stock, it seems unlikely that higher insulation levels could reduce energy demand in new industrial buildings. It has been assumed that demand for both space-heating and lighting could be reduced to 70% of the current

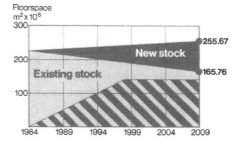

INDUSTRIAL SECTOR

Fig. 4A.1. Potential for incorporating passive solar features (industrial sector).

Table 4A.1 Passive solar potential ranking by building category (standard matrix)

| Building category — | | | | | | | Estimated total existing floor area | ×10⁶ m² | | | |

Building category —	Estimated total existing floor area	$\times 10^6$ m²
	Estimated annual energy consumption (thermal)	PJ
	Estimated annual energy consumption (lighting)	PJ

Building Type	Environment required	Building parameters						Passive solar potential ranking		Proportion of stock	Weighted rankings	
		Fabric		HVAC	Internal gains		Usage					
		Mass	Glazing		Lights	Others		Thermal	Lighting		Thermal	Lighting
1		Heavy H	<10% VL	H	<10 W/m² L	<5 W/m² L	Cont. C	0 = non-starter				
2		Medium M	10–20% L	H&V	10–25 W/m² M	5–50 W/m² M	Inter. I	10 = ideal				
3		Light L	20–30% M	A/C	>25 W/m² H	>50 W/m² H	Var. V					
⋮		Add'l Storage S	30–60% H									
			>60% VH									

Environment required: Loose Fit LF Comfort CT Close Control CC

Table 4A.2 Passive solar potential ranking by building category

Building category — INDUSTRIAL

Estimated total existing floor area	226	×10⁶ m²
Estimated annual energy consumption (thermal)	475	PJ
Estimated annual energy consumption (lighting)	84	PJ

Legend:

- Environment required: Loose Fit — LF, Comfort — CT, Close Control — CC
- Fabric — Mass: Heavy H, Medium M, Light L, Add'l Storage S
- Glazing: <10% VL, 10–20% L, 20–30% M, 30–60% H, >60% VH
- HVAC: H&V H, A/C A/C
- Internal gains — Lights: <10 W/m² L, 10–25 W/m² M, >25 W/m² H
- Internal gains — Others: <5 W/m² L, 5–50 W/m² M, >50 W/m² H
- Usage: Cont. C, Inter. I, Var. V
- Passive solar potential ranking: 0 = non-starter, 10 = ideal

Building Type	Environment required	Fabric Mass	Glazing	HVAC	Internal gains Lights	Internal gains Others	Usage	PSP ranking Thermal	PSP ranking Lighting	Proportion of stock	Weighted rankings Thermal	Weighted rankings Lighting
1	LF	L:M	VL:L	H	L	L	—	0	0	0.10	0	0
2	LF	L:M	M	H	L	L	—	0	1	0.10	0	0.10
3	LF	L:M	VL:L	H	L	M:H	—	0/1	0	0.10	0/0.1	0
4	LF	L:M	M	H	L	M:H	—	0/1	1	0.10	0/0.1	0.10
5	LF	L:M	H	H	L	L:M	—	2	1/2	0.10	0.2	0.1/0.2
6	CT	L	VL:L	H	L	M	—	1	0	0.07	0.07	0
7	CT	L	M	H	L	M	—	1	1	0.07	0.07	0.07
8	CT	L	VL:L	H	M	M:H	—	0	0/1	0.07	0	0/0.07
9	CT	M:H	M	H	M	M:H	—	1/2	3/4	0.07	0.07/0.14	0.21/0.25
10	CT	M:H	VL:L	H	M	M:H	—	1/2	0/1	0.07	0.03/0.06	0/0.07
11	CT	M:H	M	H	M	M:H	—	4	3/4	0.03	0.12	0.09/0.12
12	CT	L:M:H	H	H:H&V	L:M	M	—	2	3/4	0.03	0.06	0.09/0.12
13	CT	L:M:H	H	H:H&V	L:M	H	—	0/1	1	0.03	0/0.03	0.09/0.12
14	CT	L:M:H	VL:L	H:H&V	M:H	H	—	0/1	5/6	0.03	0/0.03	0.03
15	CT	L:M:H	M	H:H&V	M:H	H	—					0.15/0.18
										1.00	0.62/0.98	1.03/1.46

Table 4A.3 Passive solar potential ranking by building category

Building category— HOSPITALS

Estimated total existing floor area	34 ×10⁶ m²
Estimated annual energy consumption (thermal)	66 PJ
Estimated annual energy consumption (lighting)	28 PJ

Building parameters

Key:
- Environment required: Loose Fit LF; Comfort CT; Close Control CC
- Mass: Heavy H; Medium M; Light L; Add'l Storage S
- Glazing: <10% VL; 10–20% L; 20–30% M; 30–60% H; >60% VH
- HVAC: H&V H; A/C
- Internal gains Lights: <10 W/m² L; 10–25 W/m² M; >25 W/m² H
- Internal gains Others: <5 W/m² L; 5–50 W/m² M; >50 W/m² H
- Usage: Cont. C; Inter. I; Var. V
- Passive solar potential ranking: 0 = non-starter, 10 = ideal

Building Type	Environment required	Mass	Glazing	HVAC	Internal gains Lights	Internal gains Others	Usage	Passive solar potential ranking Thermal	Passive solar potential ranking Lighting	Proportion of stock	Weighted rankings Thermal	Weighted rankings Lighting
1	CT	M:H	VL	H	L:M	L	C:V	0	1	0.10	0	0.1
2	CT	M:H	L	H	L:M	L	C:V	2	2	0.10	0.2	0.2
3	CT	M:H	M	H	L	L	C:V	2	3	0.10	0.2	0.3
4	CT	M:H	H	H	M	L	C:V	5	4	0.10	0.5	0.4
5	CT	M:H	M	H	M	L	C:V	4	4	0.10	0.4	0.4
6	CT	M:H	H	H	M:H	M	C:V	4	5	0.10	0.4	0.5
7	CT	M:H	M	H	M:H	M	C:V	2	4/5	0.10	0.2	0.4/0.5
8	CC	M:H	H	H	L:M:H	L	C:V	2	6	0.10	0.2	0.6
9	CC	M:H	L	H&V:A/C	L:M	L	C:V	0	2/3	0.04	0	0.08/0.12
10	CC	M:H	M	H&V:A/C	L:M	L	C:V	3	3/4	0.04	0.12	0.12/0.16
11	CC	M:H	H	H&V:A/C	L:M	L	C:V	3	5/6	0.04	0.12	0.2/0.24
12	CC	M:H	M	H&V:A/C	L:M	M	C:V	0/1	3/4	0.04	0/0.04	0.12/0.16
13	CC	M:H	H	H&V:A/C	L:M	M	C:V	0/1	5/6	0.04	0/0.04	0.2/0.24
										1.00	2.34/2.42	3.62/3.92

Table 4A.4 Development of passive solar potential in industrial buildings, 1985–2010

Year	Existing stock		New-build stock		Total (PJ)	Total (mtce)	Cumulative Total (mtce)
	Heating (PJ)	Lighting (PJ)	Heating (PJ)	Lighting (PJ)			
1985	0.189	0.052	0.040	0.011	0.292	0.011	0.011
1986	0.378	0.104	0.080	0.022	0.584	0.022	0.033
1987	0.567	0.156	0.120	0.033	0.876	0.033	0.066
1988	0.756	0.208	0.160	0.044	1.168	0.044	0.110
1989	0.945	0.260	0.200	0.055	1.460	0.055	0.165
1990	1.134	0.312	0.240	0.066	1.752	0.066	0.231
1991	1.323	0.364	0.280	0.077	2.044	0.077	0.308
1992	1.512	0.416	0.320	0.088	2.336	0.088	0.396
1993	1.701	0.468	0.360	0.099	2.628	0.100	0.496
1994	1.890	0.520	0.400	0.110	2.920	0.111	0.607
1995	2.079	0.572	0.440	0.121	3.212	0.122	0.729
1996	2.268	0.624	0.480	0.132	3.504	0.133	0.862
1997	2.457	0.676	0.520	0.143	3.796	0.144	1.006
1998	2.457	0.676	0.560	0.154	3.847	0.146	1.152
1999	2.457	0.676	0.600	0.165	3.898	0.148	1.300
2000	2.457	0.676	0.640	0.176	3.949	0.150	1.450
2001	2.457	0.676	0.680	0.187	4.000	0.152	1.602
2002	2.457	0.676	0.720	0.198	4.051	0.153	1.755
2003	2.457	0.676	0.760	0.209	4.102	0.155	1.910
2004	2.457	0.676	0.800	0.220	4.153	0.157	2.067
2005	2.457	0.676	0.840	0.231	4.204	0.159	2.226
2006	2.457	0.676	0.880	0.242	4.255	0.161	2.387
2007	2.457	0.676	0.920	0.253	4.306	0.163	2.550
2008	2.457	0.676	0.960	0.264	4.357	0.165	2.715
2009	2.457	0.676	1.000	0.275	4.408	0.167	2.882
2010	2.457	0.676	1.040	0.286	4.459	0.169	3.051

average levels: $1.47\,GJ/m^2$ and $0.26\,GJ/m^2$ respectively.

The growth rate for new industrial buildings is assumed to be 1.5%. The annual saving in space-heating energy arising from passive solar design approaches could therefore be 0.040 PJ ($225\,000\,000\,m^2 \times 1.47\,GJ/m^2 \times 0.8\% \times 1.5\%$). The annual saving for demand in lighting energy from passive design could be 0.011 PJ ($225\,000\,000\,m^2 \times 0.26\,GJ/m^2 \times 1.25\% \times 1.5\%$).

Table 4A.4 summarises the cumulative contributions from modifications to existing and new-build industrial stock over the next 25 years. By 1997 savings will have accumulated to one million tonnes of coal equivalent; by 2004 it will be 2 mtce; and by 2010 it will be 3 mtce. Space heating will contribute almost 80% of these savings, and lighting the balance. Table 4A.5 presents an example of the cost-effectiveness of savings in a factory. Table 4A.6 provides energy and mtce savings information for hospitals over the same period.

Table 4A.5 Cost-effectiveness—factory example

Factory data

Area	$1\,000\,m^2$
Length	50 m
Width	20 m
Wall height	5 m

Thermal figures

Annual energy consumption for space heating (based on industrial average of 2.1 GJ/m² p.a.) 1 000 × 2.1 GJ	= 2 100 GJ (584 000 kWh)
Annual space heating cost (based on gas at 1.2 p/kWh, delivered)	= £7 000
If the factory has a thermal passive solar potential of 0.8% (based on the thermal weighted ranking of Table 4A.2) the annual saving could be	= £56
The amount available for investment, based on a 5-year simple payback period	= £280
If the long wall (50 × 5 = 250 m²) faces south this would allow an expenditure on passive features of 280 ÷ 250	= £1.28/m²
If the short wall (20 × 5 = 100 m²) faces south this would allow an expenditure on passive features of 280 ÷ 100	= £2.80/m²

Lighting figures

Annual energy consumption for lighting is 1 000 × 0.37 GJ	= 370 GJ (102 860 kWh)
The potential passive saving is £5 660 × 1.25%	= £70.75
The amount available for investment based on a 5-year simple payback period	= £353
If the long wall with 20% glazing faces south the permitted expenditure would be	= £7/m²
If the short wall with 20% glazing faces south the permitted expenditure would be	= £17.50/m²

Table 4A.6 Development of passive solar potential in hospital buildings, 1985–2010

Year	Existing stock		New-build stock		Total (PJ)	Total (mtce)	Cumulative Total (mtce)
	Heating (PJ)	Lighting (PJ)	Heating (PJ)	Lighting (PJ)			
1985	0.078	0.053	0.016	0.011	0.158	0.006	0.006
1986	0.156	0.106	0.032	0.022	0.316	0.012	0.018
1987	0.234	0.159	0.048	0.033	0.474	0.018	0.036
1988	0.312	0.212	0.064	0.044	0.632	0.024	0.060
1989	0.390	0.265	0.080	0.055	0.790	0.030	0.090
1990	0.468	0.318	0.096	0.066	0.948	0.036	0.126
1991	0.546	0.371	0.112	0.077	1.106	0.042	0.168
1992	0.624	0.424	0.128	0.088	1.264	0.048	0.216
1993	0.702	0.477	0.144	0.099	1.422	0.054	0.270
1994	0.780	0.530	0.160	0.110	1.580	0.060	0.330
1995	0.858	0.583	0.176	0.121	1.738	0.066	0.396
1996	0.936	0.636	0.192	0.132	1.896	0.072	0.468
1997	1.014	0.689	0.208	0.143	2.054	0.078	0.546
1998	1.014	0.689	0.224	0.154	2.081	0.079	0.625
1999	1.014	0.689	0.240	0.165	2.108	0.080	0.705
2000	1.014	0.689	0.256	0.176	2.135	0.081	0.786
2001	1.014	0.689	0.272	0.187	2.162	0.082	0.868
2002	1.014	0.689	0.288	0.198	2.189	0.083	0.951
2003	1.014	0.689	0.304	0.209	2.216	0.084	1.035
2004	1.014	0.689	0.320	0.220	2.243	0.085	1.120
2005	1.014	0.689	0.336	0.231	2.270	0.086	1.206
2006	1.014	0.689	0.352	0.242	2.297	0.087	1.293
2007	1.014	0.689	0.368	0.253	2.324	0.088	1.381
2008	1.014	0.689	0.384	0.264	2.351	0.089	1.470
2009	1.014	0.689	0.400	0.275	2.378	0.090	1.560
2010	1.014	0.689	0.416	0.286	2.405	0.091	1.651

Section 5

Housing: The Passive Solar Approach to New and Existing Domestic Buildings

Neil Milbank

Assistant Director and Head of the Energy and Environment Group, Building Research Establishment, Garston, Watford

This paper presents the work of a sub-group, dealing with domestic buildings, of the Watt Committee Working Group on Passive Solar Building Design.

Membership of Sub-group

N. O. Milbank (Chairman)

Dr D. M. L. Bartholomew
Dr J. R. Britten
Dr V. H. C. Crisp
J. Doggart
M. E. Finbow
D. L. Goodenough
E. R. Hitchin
G. K. Jackson
E. Keeble
Dr J. G. F. Littler
J. L. Meikle
W. B. Pascall
P. H. Pitt
Dr R. Wensley

For practical purposes, in the United Kingdom, passive solar design in housing is concerned with the utilisation of solar gains through glass, predominantly through windows, but to a lesser extent through conservatories. Windows, of course, have long been an important feature in all British homes, although their proportions and area vary widely. In more northern latitudes, the traditional farmhouse has very small windows, whereas in the south houses developed with larger areas of glazing, as exemplified in 'Georgian' (eighteenth century) designs. In both, it is usual for the windows to face predominantly towards the south, whereas in most Victorian (nineteenth century) and twentieth century urban developments the orientation of the houses, and thus of their windows, is much more random.

5.1 ENERGY BALANCE OF EXISTING HOUSING

It is clear, therefore, that solar energy makes a contribution to the energy balance of existing houses, though the extent of this may be something of a surprise. In fact, estimates suggest that, on average, around 15% of energy requirements in houses are derived directly from the sun (Fig. 5.1). Within the existing stock there are many variations about this average. For example, for a well-heated but uninsulated Victorian house (Fig. 5.2(a)), solar energy will contribute less than 10% of energy needs, whereas in a house insulated to 1983 Building Regulations Standards, the contribution would be nearer to 20% (Fig. 5.2(b)). With even better insulation, *the usefulness of solar energy can drop*, because these figures assume that the orientation of windows is random in the existing stock (Fig. 5.2(c)). Those houses with more south-facing glass would do better than average for solar gains: if the southerly dispositions of glass can be combined

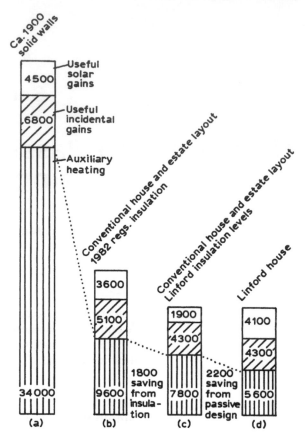

Fig. 5.2. The effect of improved insulation standards on annual space heating in houses at 19°C (kWh/year).

with even better insulation standards, solar energy can contribute up to one-third of energy needs, as found, for example, at Linford, Milton Keynes (Figure 5.2(d)).

These issues are examined in more detail here and the technical factors that determine the contribution of solar energy in a particular house are identified. This Report also explores what needs to be done to secure more widespread adoption of solar design, taking account of economic and marketing considerations.

5.2 CURRENT TECHNOLOGY: WINDOWS

The process by which solar radiation appears as useful heat in a house is quite complex. First, some of the short-wave radiation from the sun is reflected from the outer surface of the glass, some is transmitted directly into the room and some is absorbed in the glass itself. Of the absorbed energy, approximately one-third eventually appears in the room space and two-thirds is released to the outside air. A small fraction of the transmitted radiation is

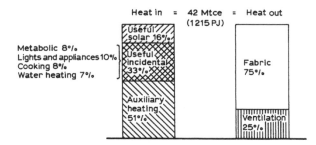

Fig. 5.1. Space heating energy balance in the existing UK housing stock.

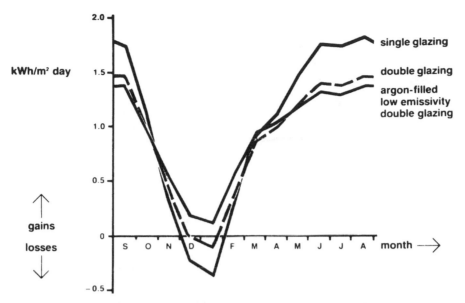

Fig. 5.3. Energy balance through unobstructed south-facing windows at Kew.

reflected out again and the remainder is subsequently absorbed by the surface of the room, the furniture and the fabric. As these surfaces heat up they release energy into the room and so create the possibility of reducing the space-heating requirement. 'Useful' solar gains are those that reduce space-heating demand, either at the time when the gain is received, or later when heat stored in the building is released. Thus a significant number of factors determine the usefulness of solar energy; they include the following:

— Latitude and weather
— Orientation and overshading
— Transmission characteristics of the glazing
— Pattern of occupancy
— Responsiveness of heating controls and heating systems
— Thermal mass of the building and contents
— Thermal insulation standards

The first three items in this list are probably the most important of these variables; they are illustrated in Fig. 5.3, which gives the monthly energy balance through south-facing glazing at Kew. For single glazing there is an overall heat loss during the months of December, January and February; for double glazing these losses are effectively eliminated; with low-emissivity double glazing there is a net heat gain at all times of the year.

Of course, not all windows can face south, nor can they be completely unobstructed. The effects of other orientations and various degrees of skyline

obstructions are shown in Fig. 5.4. From this it is clear that windows intended to provide useful solar gains should preferably be orientated within 30° of south, although orientations up to 45° on either side of south can realise about half the possible solar gain. Modest obstructions on the horizon, say up to 10°, are not too important, but by the time the obstruction angle is 25° the potential solar benefits are again reduced to less than half the unobstructed values. Conversely, solar gains may be increased by radiation reflected from the ground or adjacent walls if they are made of suitable light-coloured materials.

The opportunity to benefit from orientation and site layout has been incorporated into a number of exemplar estates, particularly at Milton Keynes and Basildon. The houses have a preponderance of glass in the living rooms and main bedrooms, which face towards the south. Kitchens and bathrooms are on the north side. The site layouts limit overshading as much as possible and incorporate house plans to suit both broad and narrow plot sizes. The heating systems are controlled by a master thermostat in the living room; it controls the boiler and there are thermostatic radiator valves in other rooms.

The monitored results from Linford (Milton Keynes) show a contribution from the sun of some 30% of space-heating requirements. This was achieved without problems of overheating; it was found that the peak temperature inside was always below the outside temperature and only some 1·5°C above that of houses of traditional design. The

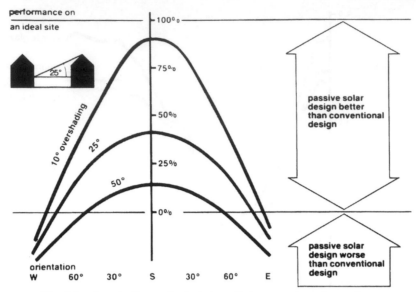

Fig. 5.4. Overshading and orientation effects on passive solar performance.

worst problems were in bedrooms, but even there the occupants found the conditions acceptable.

Studies have been made of the likely performance of the Linford house designs in other locations, using window area and type of glass as design variables. For example, if the basic house type were located at Kew, the lowest energy consumption with single glazing would be achieved if the window area were about 60% of the Linford design; whereas, with double glazing, the optimum is about the same as the value adopted for Linford. In contrast, at Lerwick the optimum size for single glazing would in fact be zero—which perhaps explains why the traditional Scottish farmhouse has such small windows. Double glazing at Lerwick would be at an optimum with about 80% of the Linford value. The optimum glazing with heat-reflecting glasses is in the range 1.0–1.4 for both sites. Overall, of course, the heating requirements are inevitably greater at Lerwick, but over the

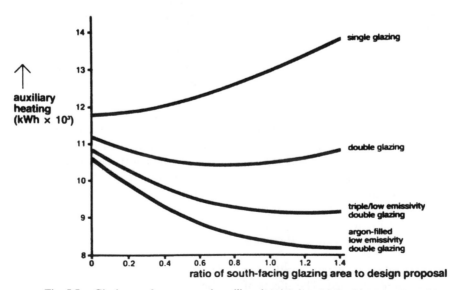

Fig. 5.5. Glazing performance and auxiliary heating in a Linford house at Lerwick.

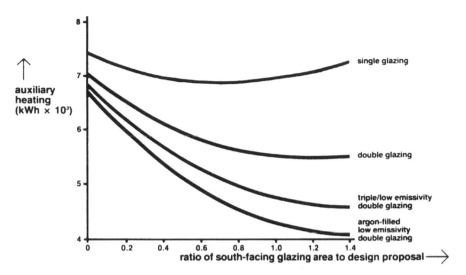

Fig. 5.6. Glazing performance and auxiliary heating in a Linford house at Kew.

course of the year the useful solar gain through double glazing is almost the same there as that realised at Kew (Figs. 5.5 and 5.6).

Clearly, orientation, window area and overshading are key issues for new housing. For existing housing, of course, these are largely predetermined, but even then, when windows are being replaced the careful choice of the type of glazing—single, double or low-emissivity—can produce more effective use of solar energy and a reduction in space-heating requirements.

5.3 CURRENT TECHNOLOGY: CONSERVATORIES

Conservatories have become popular with British house-owners as a way of providing extra floor space, although they may not be fully habitable in the depths of winter. However, a suitably placed conservatory can also offer energy-saving and comfort benefits to the main body of the house. Ways of achieving this are listed here.

(1) Conservatories trap solar gains by the same process as windows in rooms, though, of course, they have more glass than the normal room; conservatories therefore heat up and cool down much more quickly. What they provide is a space with a climate that is usually intermediate between inside and outside conditions. For example, they could provide the equivalent of April external temperatures during the months of January and February, thus reducing fabric

losses from the houses.

(2) Conservatories provide a source of warm air which can be diverted into the rest of the house and so increase the solar contribution to space heating.

Conservatories can be incorporated into the design of new housing as well as added to existing housing, and their several benefits form an attractive sales feature.

5.4 FUTURE TECHNOLOGIES

For the future, there are two main domestic areas in which improved design, from the passive solar viewpoint, can be expected: first, improvements in the performance of windows; and, second, the potential development of various forms of glazed walling.

The main improvement in the performance of windows is likely to be the development of systems that utilise solar energy selectively, i.e. that admit solar energy when it is required but reduce heat losses when the sun is not shining. In this latter context there are considerable opportunities for shutters and better curtaining systems, whether manually or automatically operated. However, there is also the opportunity to develop more sophisticated types of glass which can change their transmission characteristics. Photochromic glasses are already commonplace for non-building uses, and the development of electrochromic glasses could provide opportunities for much more sophisticated systems of automatic control—for example, relating the glass transmission characteristics to internal temperature.

There have been experiments on glazed walls in many parts of the world, but in the UK experience to date is limited. The largest experiment is with houses at Bebbington, where a number of bungalows use fan systems to transport the warm air from the glazed wall to the bedrooms on the north-facing side of the house. These seem to work effectively as far as the occupants are concerned, but the costs of such systems at present cannot be justified in terms of the energy saved. Other experiments are under way on smaller solar collectors which could be attached to the walls of existing houses and transfer warm air into the adjacent rooms either by means of a fan or by natural convection. It is too early yet to know how effective these will be, but clearly they could become attractive, particularly on existing houses which have large areas of south-facing walls.

Glazed roof spaces are also a possibility, but in order to get the heat down to other parts of the building they require some mechanical components. They thus become a hybrid between active and passive systems.

5.5 BARRIERS TO IMPLEMENTATION

In an ideal world, market forces could be expected to lead to the incorporation of passive solar technology into housing in a very short period of time. However, the building industry—particularly the housing sector—is usually cautious when considering new concepts and ideas. In part this is because many parties are involved in developing a new house design or altering an existing one. Rarely are the householder and his contractor the only parties to a contract. More usually there are financial organisations (banks or building societies), building control and regulations (which ultimately involve administrators and politicians), professional advisers, manufacturers, fuel interests; each has an important part to play in providing the information needed to reach a satisfactory outcome. Unless each of these promotes more or less the same message at the same time, the result is confusion—which leads to a lack of confidence and finally to an unwillingness to change from proven practice.

What then are the major issues that should be addressed, so helping to establish confidence amongst this wide range of interested parties? Not surprisingly, they all concern questions of benefits, costs and risks.

5.5.1 Design techniques

First is the question of design techniques. At present, when a house design is developed, it is rare for significant professional effort to be put into the analysis and specification of environmental and energy performance. Currently, passive solar design would seem possible, to an outsider, only with complicated computer-modelling techniques. What is needed is the development of a whole series of design appraisal methods, ranging from the simplest to the most complicated, so that an appropriate model can be chosen to resolve the particular design question as it arises. For instance, it should be possible to develop simple rules of thumb; e.g. the total area of south-facing glass should not exceed a proportion—say one-third—of floor area. Such a figure would need to be developed from more sophisticated models and should take into account not only the value of solar energy but also the risk of overheating. It is also important to ensure that, whatever models are derived, they all have a reasonable relationship to each other so that similar solutions are realised whichever design technique is used.

5.5.2 Performance data

In practice, there is little hard evidence of the performance of occupied houses that incorporate solar features. This should not be a surprise, since it takes a considerable time to develop and build suitable designs, then to monitor the houses while occupied and finally to analyse the results in a meaningful and unambiguous way so that they have a real effect on future designs. Inevitably, because of the influence of housholders, field trials in occupied housing show a range of responses and therefore can lead to a number of interpretations. For example, the present studies have drawn extensively from results for the Linford estate; yet, in different published reports, the average savings quoted vary from 1400 to 2200 kWh per year. This is a very significant difference and results from adjustments to allow for variations in weather, user behaviour, 'clutter' obstructing sunlight that would fall on windows, etc. Less ambiguous presentations are needed if confidence in the benefits of passive solar design is to be secured.

An important factor that affects site layout, and hence the opportunities to optimise orientation and reduce overshading, is the *sequence* in which design decisions are taken. The layout of roads and plots is usually a very early decision in the development of a site. Typically it is based on minimising the lengths of road construction and mains services (gas, electricity, drainage mains, etc.) and may be decided without reference to the detailed design of the house. For a passive solar site, design, road layout, plot layout and house design need to be considered together so that the energy and environmental benefits of good solar access are properly taken into account.

A further issue is non-familiarity with the *design concept*. This aspect includes terminology. It raises the question of whether 'passive solar' is the correct term when, for practical purposes, the issue is the siting of windows and the provision of conservatories in well-insulated houses. Most people recognise windows and conservatories—few would recognise that this is what passive solar is about.

5.5.3 Perceived value

The next important issue is less a technical than a marketing one: the identification of some *perceived value* of the feature to help sell the house. It is common ground that the provision of fitted kitchens is an important consideration in the sale of new and existing housing. However, it has not yet proved possible to market solar features in a similar way, although research already done for the Department of Energy has found that at present some estate agents do emphasise south-facing aspects, etc. Somehow this initiative has to be extended to identify the related energy benefits of solar features if they are to figure large in the decision to buy a house.

5.5.4 Costs and benefits

The final concern is *costs and benefits*. Again, work for the Department of Energy has thrown light on this issue (Fig. 5.7) by showing the cost of installation and the saving in energy bills. Back-up work shows that there are a few households for whom saving money is of little importance—for them the prestige of being amongst the first into the field is compensation, whatever the energy savings. For others, the public sector rate of return is sufficient; and for others still, there is no interest in reducing energy bills unless the added cost is close to zero. These groups are identified by the three radiating bands shown in Fig. 5.7. Also shown are the costs and savings related to two measures, one a conservatory, the other the change to south-facing glass. The costs and savings for both measures are presen-

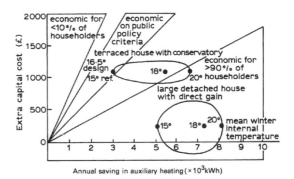

Fig. 5.7. Comparison of capital costs and annual energy savings.

ted in the form of an ellipse to show the uncertainty both in predicting energy saving and in determining the marginal cost of a change in building design.

5.6 ENERGY SAVINGS AND ENVIRONMENTAL BENEFITS

What has been said identifies the barriers: it remains to restate the potential benefit—that is, that valuable energy savings and environmental benefits can be achieved in housing if more thought is given to site layout, orientation and choice of glazing. The target could well be a Linford type of design wherein the solar energy contribution can be about one-third of space-heating requirements, if the rest of the house is well insulated and the heating well controlled. If this were achieved in all new houses, it would make a noticeable contribution to the reduction of energy demand.*

APPENDIX 5A: THE POTENTIAL BENEFIT NATIONALLY

New estimates of energy-saving potential arising from passive solar housing estate layout are given in the report, *A study of passive solar housing estate layout* (ETSU S-1126, AERE, Harwell, April 1986), and are based on figures which take account of house type, site density, climatic region, housebuilding activity and the anticipated speed of take-up of these design methods. These estimates deal with not only the annual energy saving in 20 years' time but also the accumulated savings for the whole of the 20-year period. They are more realistic because they take account of the less than ideal conditions that

---

* The potential benefit nationally was the subject of a brief paper, presented to the Watt Committee working group by M. E. Finbow, based on the ECD report *Passive solar housing in the UK*. More recent results are reproduced as Appendix 5A.

exist on real sites, such as density and obstruction.

The annual primary energy saving after 20 years following the gradual introduction of passive solar house design and layout principles into future housebuilding is estimated to be 0.2 mtce, the cumulative saving over the next 20 years being about 1.5 mtce.

These estimates have been built up taking into account the following factors:

— the effect of site density on passive solar benefit;
— the proportion of each house type that is likely to achieve full benefit from passive solar design;
— corrections for climatic zone and unfavourable site conditions;
— the potential energy saving from solar dwelling design;
— estimates of future housebuilding activity;
— market take-up of passive solar design principles.

By applying our conclusions about site density to the regional housebuilding statistics given in the Appendix,* we have estimated the number of potential solar houses that could be built annually, given certain predictions about levels of future housebuilding activity. To these estimates was applied a correction for market take-up.

Full details are given in the Appendix,* but the main assumptions made in building up the estimates are as follows:

(1) Due to the nature of their sites (density, etc.), 100% of detached houses, bungalows and semi-detached houses are able to achieve full solar benefit, but only 50% of terraced houses and 25% of flats.

(2) For the first 5 years, annual housebuilding is taken to be 160 000 (private sector) and 40 000 (public sector), reducing thereafter to 125 000 (private) and 25 000 (public).

(3) The take-up of passive solar principles in the private sector will lag slightly behind the public sector.

(4) Annual energy savings from the building of passive solar rather than conventional houses is assumed to be 1000 kWh, the equivalent saving for flats being 500 kWh; in both cases the savings are only for dwellings in suitable locations on favourable sites.

* This refers to the Appendix to the report quoted.

Table 5A.1 Annual energy savings nationally after 20 years

Sector	Primary energy savings nationally (mtce)			Share of total (%)
	Houses	Flats	Total	
Public	0·029	0·004 0	0·033 0	16
Private	0·165	0·004 6	0·169 6	84
Total	0·194	0·008 6	0·202 6	100

Table 5A.2 Total energy savings nationally over the next 20 years

Sector	Primary energy savings nationally (mtce)			Share of total (%)
	Houses	Flats	Total	
Public	0·235	0·033	0·268	18
Private	1·200	0·033	1·233	82
Total	1·435	0·066	1·501	100

Two estimates are given of the effect nationally:

— annual energy savings at the end of the next 20 years;
— total cumulative energy savings achieved over the 20-year period.

The build-up of the figures is given in the Appendix,* but Tables 5A.1 and 5A.2 give a simple breakdown of the totals in terms of primary energy between:

— public and private sector;
— houses and flats.

All the savings are based on the calculation that over the next 20 years a total of 616 000 solar houses and 56 000 solar flats will have been constructed.

Clearly, private sector houses offer the greatest potential for energy saving by passive solar design, contributing about 80% to the total. The savings from flats account for less than 5% of the total.

The only other source against which to compare our estimates is the report *Passive Solar Design in the UK*, prepared in 1980 by Energy Conscious Design. Their figure for annual energy savings from passive solar house design at the end of a 20-year period convert to a primary energy saving of 0.39 mtce, compared with our estimate of 0.2 mtce. This is due not to differences in our assumptions about housebuilding activity and market take-up, which are very similar, but to the fact that we have made further corrections for site density, site suitability and the reduced solar altitude in northern latitudes.

Section 6

Conclusions and Recommendations

Patrick O'Sullivan

This paper has been prepared by the Chairman of the Watt Committee Working Group on Passive Solar Building Design, taking account of the proceedings at the Eighteenth Consultative Council Meeting of the Watt Committee on Energy, held in London on 23 October 1985, the contents of Sections 2–5, and discussion at the final meeting of the Working Group.

The papers in this Report indicate that the benefits that accrue to a building and its occupants from a proper consideration of solar radiation are greatest when the so-called 'passive solar component' is seen in perspective, as a natural part of an integrated approach to climatically interactive (as opposed to climatically rejecting) low-energy building design. In this way, 'passive solar' has come of age.

6.1 CLIMATICALLY INTERACTIVE BUILDING DESIGN

The debate therefore is now no longer whether such climatically interactive buildings can or in fact should be built, but is rather concerned with the two following matters:

(1) The amount of solid evidence available to encourage and reassure clients who wish to request such buildings, designers who wish to design them and tenants/owners who wish to buy them.

(2) The availability of evidence in the proper form to provide the best advice and guidance for the wide variety of people interested in climatically interactive building solutions.

The first concern of the Working Group, however, is to review or define what we understand by the term a 'climatically interactive' design or indeed a 'climatically interactive' solution. It is a design that demonstrates the ability to integrate the scope to use the location, form and fabric of the building to admit, store or, at times, reject and distribute solar energy within that building so as to improve its thermal and visual environment. It is therefore but one small step further down a well understood and well trodden road. A climatically interactive solution, therefore, is a building solution that achieves, at least in some part, these ends.

'Success' we would define as a mixture of improvements in amenity (people appear to like interactive building solutions), often at no extra constructional or maintenance costs, with the financial savings in running cost achieved as a result of the solar displaced fuel.

The reality may well be that the savings may not be as great as we would like or predict, because of the difference between design and constructed standards, for which there is an increasing body of evidence. This difference may well mask the true potential/value of the solar displaced fuel component.

6.2 AVAILABLE KNOWLEDGE

Turning now to the papers in this Report, I believe they show that a great deal of knowledge and evidence is available, perhaps more than was previously thought. This knowledge and experience is greatest in the domestic and educational sectors, less in the commercial and retail sector, and least in the industrial and other buildings sectors.

Our knowledge to date is largely based on the communal experience of individually built exemplars rather than on any structured programme of measurement and evaluation which is now, however, in hand. But we are in no doubt that the era of the climatically interactive building as a viable solution has arrived.

Our weakness still lies in our ability to predict. Most *design* models currently available to us handle the climatically interactive effects of solar radiation inadequately at anything other than the domestic level. For the prediction of the advantages of integrating climatically interactive procedures into the designs of our schools and factories (i.e. simple large building spaces), we have to turn to our larger, predominantly research-based, models. Complex buildings, like large offices, large hotels, retail premises and hospitals cannot, we believe, be adequately handled yet, though basic principles concerning, for instance, screening, shading and overhangs are clear.

6.3 DOMESTIC PREMISES

In the domestic sector, we have the largest number of built and occupied examples of climatically interactive design. Moreover, the majority of these houses are liked by their occupants. We can with some confidence, therefore, encourage others to go ahead and build.

The built examples encompass a whole range of interactive solutions that aim to improve amenity and to optimise the use of solar displaced fuel in the spring and autumn. They avoid overheating in summer and are energy-efficient in winter.

The need is to develop methods and techniques that will enable us to evaluate a reputable sample of these houses to see which solutions, if any, work best, and in particular if any combination of building components produces better building solutions than others. For example, do particular solutions offer more opportunities than others, bearing in mind the reality of the site limitations of density, overshadowing, orientation, etc.? Does any particular fabric solution interact with heating/lighting systems in a particularly efficient and pleasing way? Are there any fabric/system design interactions that fit the normal range of occupancy patterns better than others? Does an integrated systems approach to climatically interactive design produce better results, according to the criteria given above, than an 'add on a solar feature' non-integrated approach?

In other words, is there a best buy or buys, and are particular solutions to be avoided? More importantly, perhaps, how do the best climatically interactive solutions compare with the best climatically rejecting housebuilding solutions?

We need standard, simple, robust, fast, credible, economic evaluation techniques. In addition, we need new means of presenting the advantages of climatically interactive design to the wide variety of interests involved and of getting these advantages accepted and used.

6.4 EDUCATIONAL PREMISES

Our knowledge base for schools is of a similar size and state to that for our houses. We have many good exemplars of climatically interactive design that are educationally pleasing and efficient, economical to run and pleasant to work in. We can with confidence encourage others to build.

The need, again, is similar: to develop methods and techniques in order to evaluate what we have, to determine the most favourable options—the range of 'best buys'.

6.5 COMMERCIAL, RETAIL, AND OTHER BUILDINGS SECTOR

In this sector, once again the aim is to balance the advantages of daylighting against the disadvantages of overheating while preserving an energy-efficient construction. In other words, is it possible to save on direct-acting electrical lighting load while at the same time avoiding solar overheating and minimising the cooling load, if any?—all this in buildings that the occupants prefer.

Although the principles are established and both naturally ventilated and air-conditioned solutions articulated, we have fewer examples.

The need is again to develop methods and techniques and to evaluate those examples that we possess in order to determine, if possible, a range of best buys. This is no easy matter in complex commercial buildings, as it is in this and the industrial and other buildings sector that our methodologies and our models are least well developed.

We need to build more exemplars. There is encouraging evidence, albeit of an anecdotal form. Moreover, the main texts show that we can offer a clear exposition of the known rules.

Although examples of good climatically interactive solar design exist in the retail sector, they are, again, few and far between, and in the other building sectors, particularly the industrial sector, they are, to the best of our current knowledge, almost non-existent in the United Kingdom.

The potential appears large. The actual experience, however, is small. Overall, the need is to teach and explain the rules, so as to encourage exemplars. Again, the anecdotal evidence that we have is encouraging; we cannot say more. The rules are known and have regard to suitability for purposes, constructional costs, flexibility and energy efficiency.

6.6 OVERALL CONCLUSIONS

The way forward is to encourage more climatically interactive buildings by the presentation of best-buy documented exemplars. This, we believe, can now be demonstrated in the domestic and educational sectors. There is some evidence to encourage building in the commercial and retail sectors, but very little, if any, in the UK for the other building types.

Elsewhere in this Report the results of consideration of a simple methodology for evaluating 'passive' buildings and a catalogue of such buildings are described. A generally applicable technique is required. Overall, we need, therefore:

(1) A catalogue of passive buildings, by type, and how they work.

(2) A methodology for evaluating the buildings that we do have in order to determine the best buys in both the domestic and particularly the non-domestic fields.

(3) A range of methods of reporting that address different audiences in different ways—e.g. case studies, exemplars, etc.

(4) A need to publicise and encourage the awareness of these viable alternatives through the media.

(5) In the commercial and other building sectors, to encourage more exemplars based on a clear exposition of the known rules, as supported by the anecdotal evidence, as a part of an integrated systems approach to design.

(6) In parallel with this, in all sectors, to produce the measured data so necessary for improving our predictive techniques—as is being done by at least one British institution. (With due modesty, I have to acknowledge that I have in mind the Department of Energy EPA Project, conducted jointly by Databuild and UWIST.)

Appendix 1

Eighteenth Consultative Council Meeting of The Watt Committee on Energy

PASSIVE SOLAR ENERGY IN BUILDINGS: REALISING THE OPPORTUNITIES

On 23 October 1985 the Watt Committee on Energy held the eighteenth in the series of its Consultative Council meetings. The theme was 'Passive Solar Energy in Buildings', and the meeting was held at the Café Royal, Regent Street, London W1. Those present were the secretaries and appointed represen tatives of the member institutions of the Watt Committee and others with professional interests in the subject of the meeting.

The meeting opened with an address by Mr David Hunt, MP, Parliamentary Under-Secretary of State at the Department of Energy. Particulars of the programme of the meeting are given below. Papers were presented informally by the chairman of the four sub-groups into which the Working Group had divided itself. There were several periods of discussion.

The papers presented at the meeting were early versions of those published in this Report; as explained elsewhere (page vi), after the meeting much additional work was carried out by the Working Group before the papers were felt to satisfy its objectives. Points that arose in discussion at the meeting were considered by the Working Group and taken into account in the preparation of the final version of the papers.

As Chairman of the Working Group, Professor P.E. O'Sullivan prepared a statement of its conclusions and recommendations which appear in Section 6 of the Report (page 55).

Programme

Passive Solar Techniques and their Application in Existing and New Technology
Prof. P.E. O'Sullivan (Chairman, Watt Committee on Energy Working Group on Passive Solar Building Design; Welsh School of Architecture, UWIST)

Opportunities for Use of Passive Solar Energy in Educational Buildings
D.M. Curtis (Essex County Council, Chelmsford)

Use of Passive Solar Energy in Offices
J. Campbell (Ove Arup Partnership, London)

Industrial, Retail and Service Buildings—Options for Passive Solar Design
W.B. Pascall (Royal Institute of British Architects)

Housing: The Passive Solar Approach to New and Existing Domestic Buildings
N.O. Milbank (Building Research Establishment, Garston)

Conclusions and Recommendations.
Prof. O'Sullivan

Contributors to discussion

Dr D.M.L. Bartholomew, Energy Technology Support Unit
G.D. Braham, The Electricity Council
H. Brown, Institution of Plant Engineers

J.R. Butlin, Building Research Establishment
Dr A.A.L. Challis, Plastics and Rubber Institute
J.S. Dodds, Property Services Agency
J. Doggart, Energy Conscious Design
J.V. Fearnley, Royal Institute of British Architects
D. Fitzgerald, Chartered Institute of Building Services Engineers
S.C. Fuller, Milton Keynes Development Corporation

K.R.G. Gray, Royal Institute of British Architects
R.S. Hackett, Institution of Gas Engineers
P.J. Hales, Anglia Building Society
Dr J.C. McVeigh, Institution of Production Engineers
Prof. J.K. Page, University of Sheffield
P.J. Webster, Institute of Physics

THE WATT COMMITTEE ON ENERGY

Objectives, Historical Background and Current Programme

1. The objectives of The Watt Commitee on Energy are:

   (a) to promote and assist research and development and other scientific or technical work concerning all aspects of energy;
   (b) to disseminate knowledge generally concerning energy;
   (c) to promote the formation of informed opinion on matters concerned with energy;
   (d) to encourage constructive analysis of questions concerning energy as an aid to strategic planning for the benefit of the public at large.

2. The concept of the Watt Committee as a channel for discussion of questions concerning energy in the professional institutions was suggested by Sir William Hawthorne in response to the energy price 'shocks' of 1973/74. The Watt Committee's first meeting was held in 1976, it became a company limited by guarantee in 1978 and a registered charity in 1980. The name 'Watt Committee' commemorates James Watt (1736–1819), the great pioneer of the steam engine and of the conversion of heat to power.

3. The members of the Watt Committee are 61 British professional institutions. It is run by an Executive on which all member institutions are represented on a rota basis. It is an independent voluntary body, and through its member institutions represents some 500 000 professionally qualified people in a wide range of disciplines.

4. The following are the main aims of the Watt Committee:

   (a) To make practical use of the skills and knowledge available in the member institutions for the improvement of the human condition by means of the rational use of energy;
   (b) to study the winning, conversion, transmission and utilisation of energy, concentrating on the United Kingdom but recognising overseas implications;
   (c) to contribute to the formulation of national energy policies;
   (d) to identify particular topics for study and to appoint qualified persons to conduct such studies;
   (e) to organise conferences and meetings for discussion of matters concerning energy as a means of encouraging discussion by the member institutions and the public at large;
   (f) to publish reports on matters concerning energy;
   (g) to state the considered views of the Watt Committee on aspects of energy from time to time for promulgation to the member institutions, central and local government, commerce, industry and the general public as contributions to public debate;
   (h) to collaborate with member institutions and other bodies for the foregoing purposes both to avoid overlapping and to maximise cooperation.

5. Reports have been published on a number of topics of public interest. Notable among these are *The Rational Use of Energy* (an expression which the Watt Committee has always preferred to 'energy conservation' or 'energy efficiency'), *Energy Development and Land in the United Kingdom*, *Energy Education Requirements and Availability*, *Nuclear Energy* and *Acid Rain*. Others are in preparation.

6. Those who serve on the Executive, working groups and sub-committees or who contribute in any way to the Watt Committee's activities do so in their independent personal capacities without remuneration to assist with these objectives.

7. The Watt Committee's activities are co-ordinated by a small permanent secretariat. Its income is generated by its activities and supplemented by grants by public, charitable, industrial and commercial sponsors.

8. The latest Annual Report and a copy of the Memorandum and Articles of Association of The Watt Committee on Energy may be obtained on application to the Secretary.

Enquiries to:
The Information Officer
The Watt Committee on Energy,
Savoy Hill House,
Savoy Hill, London WC2R 0BU
Telephone: 01–379 6875

Member Institutions of The Watt Committee on Energy

British Association for the Advancement of Science

* British Nuclear Energy Society

British Wind Energy Association

Chartered Institute of Building

Chartered Institute of Building Services Engineers

* Chartered Institute of Management Accountants

Chartered Institute of Transport

Combustion Institute (British Section)

Geological Society of London

Hotel Catering and Institutional Management Association

Institute of Biology

Institute of British Foundrymen

Institute of Ceramics

Institute of Chartered Foresters

* Institute of Energy

Institute of Home Economics

Institute of Hospital Engineering

Institute of Internal Auditors (United Kingdom Chapter)

Institute of Management Services

Institute of Marine Engineers

Institute of Mathematics and its Applications

Institute of Metals

* Institute of Petroleum

Institute of Physics

Institute of Purchasing and Supply

Institute of Refrigeration

Institute of Wastes Management

Institution of Agricultural Engineers

* Institution of Chemical Engineers

* Institution of Civil Engineers

Institution of Electrical and Electronics Incorporated Engineers

* Institution of Electrical Engineers

Institution of Electronic and Radio Engineers

Institution of Engineering Designers

* Institution of Gas Engineers

Institution of Geologists

* Institution of Mechanical Engineers

Institution of Mining and Metallurgy

Institution of Mining Engineers

Institution of Nuclear Engineers

* Institution of Plant Engineers

Institution of Production Engineers

Institution of Structural Engineers

International Solar Energy Society—UK Section

Operation Research Society

Plastics and Rubber Institute

Royal Aeronautical Society

Royal Geographical Society

* Royal Institute of British Architects

Royal Institution

Royal Institution of Chartered Surveyors

Royal Institution of Naval Architects

Royal Meteorological Society

Royal Society of Arts

* Royal Society of Chemistry

Royal Society of Health

Royal Town Planning Institute

* Society of Business Economists

Society of Chemical Industry

Society of Dyers and Colourists

Textile Institute

* Denotes permanent members of the Watt Committee Executive

Watt Committee Reports

The following Reports are available:

2. Deployment of National Resources in the Provision of Energy in the UK
3. The Rational Use of Energy
4. Energy Development and Land in the United Kingdom
5. Energy from the Biomass
6. Evaluation of Energy Use
7. Towards an Energy Policy for Transport
8. Energy Education Requirements and Availability
9. Assessment of Energy Resources
10. Factors Determining Energy Costs and an Introduction to the Influence of Electronics
11. The European Energy Scene
13. Nuclear Energy: a Professional Assessment
14. Acid Rain
15. Small-Scale Hydro-Power
17. Passive Solar Energy in Buildings
18. Air Pollution, Acid Rain and the Environment
19. The Chernobyl Accident and its Implications for the United Kingdom

For further information and to place orders, please write to:

ELSEVIER SCIENCE PUBLISHERS
Crown House, Linton Road, Barking, Essex IG11 8JU, UK

Customers in North America should write to:

ELSEVIER SCIENCE PUBLISHING CO., INC.
P.O. Box 1663, Grand Central Station, New York, NY 10163, USA

Index